# The Forest Industries Sector

## An Operational Strategy for Developing Countries

# INDUSTRY AND ENERGY SERIES

This series is sponsored by the Industry and Energy Department of the World Bank's Policy, Planning, and Research Staff to provide guidance on technical issues to government officials and World Bank staff and consultants involved in the various industrial and energy sectors and subsectors.

Other volumes to be published shortly in this series include *Proposals for Monitoring the Performance of Electric Utilities* and *The New Face of the World Petrochemical Sector: Implications for Developing Countries*.

WORLD BANK TECHNICAL PAPER NUMBER 83
INDUSTRY AND ENERGY SERIES

# The Forest Industries Sector

## An Operational Strategy for Developing Countries

Andrew J. Ewing and Raymond Chalk

The World Bank
Washington, D.C.

Second printing September 1989

Technical Papers are not formal publications of the World Bank, and are circulated to encourage discussion and comment and to communicate the results of the Bank's work quickly to the development community; citation and the use of these papers should take account of their provisional character. The findings, interpretations, and conclusions expressed in this paper are entirely those of the author(s) and should not be attributed in any manner to the World Bank, to its affiliated organizations, or to members of its Board of Executive Directors or the countries they represent. Any maps that accompany the text have been prepared solely for the convenience of readers; the designations and presentation of material in them do not imply the expression of any opinion whatsoever on the part of the World Bank, its affiliates, or its Board or member countries concerning the legal status of any country, territory, city, or area or of the authorities thereof or concerning the delimitation of its boundaries or its national affiliation.

Because of the informality and to present the results of research with the least possible delay, the typescript has not been prepared in accordance with the procedures appropriate to formal printed texts, and the World Bank accepts no responsibility for errors.



The most recent World Bank publications are described in the catalog *New Publications*, a new edition of which is issued in the spring and fall of each year. The complete backlist of publications is shown in the annual *Index of Publications*, which contains an alphabetical title list and indexes of subjects, authors, and countries and regions; it is of value principally to libraries and institutional purchasers. The latest edition of each of these is available free of charge from the Publications Sales Unit, Department F, The World Bank, 1818 H Street, N.W., Washington, D.C. 20433, U.S.A., or from Publications, The World Bank, 66, avenue d'Iéna, 75116 Paris, France.

Andrew J. Ewing and Raymond Chalk are senior industrial specialists in the Industry and Energy Division, Technical Department, of the World Bank's Europe, Middle East, and North Africa Regional Office.

**Library of Congress Cataloging-in-Publication Data**

Ewing, Andrew J., 1942-
The forest industries sector.

(World Bank technical paper, ISSN 0253-7494 ; no. 83)
1. Forest products industry. 2. Forest products industry--Developing countries. I. Chalk, Raymond, 1931- . II. Title. III. Series.
HD9750.5.E93 1988 338.1'7498'091724 88-14434
ISBN 0-8213-1072-0

## ABSTRACT

The world's forest industry has undergone important changes over the past 10-15 years as demand growth in the industrial countries has slowed, and as new, competitive producers in developing countries have successfully entered world markets. Demand will continue to grow, particularly in the developing countries, requiring increased log production and major investments in new production capacity. The future will also see increased trade in most forest products.

The purpose of this paper is to quantify these trends, in order to provide a current, overall perspective of the forest industries sector. This perspective should first, allow a realistic assessment to be made of the potential for individual countries to assume a greater role in supplying future demand, and second, assist in identifying those policies which would most effectively realize this potential.

The paper focuses primarily on recent structural and technological developments in the sector including: (i) the changes in demand growth patterns, from which medium-term projections of demand are developed; (ii) resource constraints; (iii) international comparative advantage in the production of forest products; (iv) policy issues; and (v) supply scenarios to the 1990s, based on projected demand, resource availability, comparative advantage and policy considerations.

## ACKNOWLEDGEMENT

The comparative cost data in Chapter IV were developed with the assistance of Mr. J. Easton, consultant.

# TABLE OF CONTENTS

# I. SUMMARY

The world's forest industry has undergone important changes over the past 10-15 years as demand growth in the industrial countries has slowed and as new, competitive producers in developing countries have successfully entered world markets. Demand will continue to grow, particularly in the developing countries, requiring increased log production and major investments in new production capacity. The future will also see increased trade in most forest products.

The purpose of this paper is to quantify these trends, in order to provide a current, overall perspective of the forest industries sector. This perspective should first, allow a realistic assessment to be made of the potential for individual countries to assume a greater role in supplying future demand, and second, assist in identifying those policies which would most effectively realize this potential.

## Global Demand for Forest Products

Demand for all primary forest products has increased over the past thirty years, although trends are somewhat different for each of the three primary groups of forest product. Sawnwood demand has generally grown at less than the overall rate of economic growth, and over the past decade, demand has actually contracted in some industrial countries. The demand for wood-based panels grew more quickly than the rate of economic expansion through the 1950s and 1960s, but overall growth rates now approximate economic growth. Paper demand has grown at about the rate of economic expansion since the mid-1950s.

Projections of consumption of forest products have been developed by projecting economic growth and income elasticities of demand. The median projections are as follows:

| | Actual | Projected | |
|---|---|---|---|
| | 1984-86 Average | 1995 | 2000 |
| Sawnwood (million cu m) | 463 | 509 | 538 |
| Wood-based panels (million cu m) | 111 | 138 | 156 |
| Paper & paperboard (million tons) | 195 | 257 | 299 |

These projections imply annual average growth rates of 1.0% for sawnwood, 2.3% for wood-based panels, and 2.9% for paper and paperboard. Particularly for the mechanical wood products (sawnwood and panels), these projections are less optimistic than most others which have appeared in recent years. This is a reflection of the conclusion that, at least in the industrial countries, the markets for sawnwood and plywood are mature, and there will be little further growth in demand.

To build the new capacity to provide this additional production will require an average annual investment of some US$15 billion, of which about 90% will be for pulp and paper.

## The Forest Resource

The world's commercial forests cover some 2,200 million hectares, about half of which are in developing countries and the balance in industrial countries. The area in industrial countries has stabilized, but developing countries are losing about 8 million hectares of forest each year, primarily to agricultural encroachment.

Total world log production is around 3,000 million cubic meters annually, of which slightly more than half is fuel wood. By even the most conservative estimate, the global rate of removal is less than the rate of new growth. However, in some regions of Africa and Asia, there are critical local shortages of wood, particularly fuel wood.

For expansion of forest industries, by 2000, an additional 428 million cubic meters of industrial roundwood will be required each year. There is unlikely to be any shortage of wood during this time frame: by 2000, industrial countries alone could deliver another 390 million cubic meters annually, and developing countries a further 185 million cubic meters, all on the basis of sustained yield. Under these circumstances, any country which intends to create new forest resources to compete in world markets had better be sure that it will be competitive. Even to meet domestic demand, the establishment of new forests may not be economic in comparison to purchasing wood or wood products from other efficient producers.

## Global Competitiveness and Trade

World trade in forest products is of the order of US$50 billion annually, or about 3% of total world trade. Only about US$8 billion of this total originates in developing countries, so that if trade is a reasonable indicator of historical comparative advantage, the advantage is very substantially with the industrial countries.

Developing countries do have a comparative advantage for forest products based on tropical hardwoods. There is scope to further capitalize on this advantage by further restricting the export of unprocessed tropical hardwood logs, and expanding domestic processing.

For other types of forest product, the advantage of developing countries lies in their ability to produce wood quickly, and at about one third of the cost of the industrial countries. However, true comparative advantage exists only where: (i) wood is a major component of total product cost; and (ii) the quality of the plantation wood is not a barrier to market penetration. Many forest products do not meet these criteria. A comparison of production costs for several important types of product suggests that pulpwood and pulp may be among the few which do.

Forest products prices fluctuate dramatically. Between 1976 and 1980, for example, real prices dropped by about 40-45% and are only now in the late 1980s recovering to 1976 levels. Based on long-term prices into the UK, a major importer of forest products from a variety of sources, there has been a gradual upward trend in real prices of 1-2% per year for most major forest products. This reflects, but does not fully compensate for, increased real costs for wood, energy and pollution abatement. Particularly for pulp and paper, only by building larger and more efficient plants has it been possible to maintain the growth in real prices at these low levels.

## Policy Issues

On the resource side, consideration must be given to policies with respect to forest ownership, planting or replanting, and management of the natural forests. The objective of such policies should be to maximize the benefits of utilizing the forests while maintaining acceptable environmental criteria, and giving due consideration to other legitimate uses. To best deal with the conflict between the utilization objective and the protection caveats, it is often appropriate that ownership of forests is concentrated in the public sector. However, recognition must be given to the fact that security of wood supply is an essential prerequisite to the establishment of large-scale wood-using industries.

Coordination between forestry and industry is essential, to ensure that adequate forests are in place to supply industries, and viable industries in place to use the output from industrial forests. While apparently self-evident, there are enough expensive examples of the effects of ignoring this policy to warrant its repetition here.

On the industry side, many of the policy issues are rather similar for forest industries as for other industries, with the most important issues relating to trade and to exchange rates. Trade policies must be such as to stimulate the development of efficient forest industries to supply domestic markets and, where appropriate, to export. If "infant industry" protection is deemed necessary, it should be eliminated after a fixed period of time. For exporting industries, the forming of groups for marketing is often appropriate, but participation in such groups must be voluntary to ensure that the system remains efficient.

Exchange rates should be set and adjusted to ensure that importers and exporters are paying prices or receiving revenues which truly reflect the international value of the goods they are trading. In countries with foreign exchange controls, industries with a proven capacity to generate export earnings must have access to enough of their foreign exchange earnings to enable them to import the necessary supplies, spare parts and other inputs required to remain competitive in world markets.

## Supply Scenarios for the 1990s

Although resource availability will not constrain the world-wide supply of industrial wood products, Western Europe and Japan will continue to be major deficit areas and will be joined by China and the other developing countries of Asia.

The tropical hardwood market will continue much as at present. Southeast Asia will continue to dominate the supply, and the ongoing trend towards exporting sawnwood and panels rather than logs will continue.

Sawnwood in most regions will continue to be supplied primarily from internal production. The biggest new sawnwood markets for the non-traditional suppliers of Latin America are China and the other wood-deficit Asian countries. Panels will largely be supplied by increases in regional capacity. For pulp and paper, the main trade opportunity for developing countries is expansion in shipments of pulpwood and pulp.

Overall, the contribution of developing countries to world production of forest products will increase from about 18% at present to around 24% by the year 2000, and their share of trade from 15% to 20%. To achieve this will require investments in forest industries in developing countries of approximately US$5.2 billion over the next 15 years, of which US$0.2 billion will be in Africa, US$3.2 billion in Asia, and US$1.8 billion in Latin America.

## II. GLOBAL DEMAND FOR FOREST PRODUCTS: TRENDS AND PROSPECTS

The three principal types of primary industrial forest products are: sawnwood, wood-based panels (plywood, particleboard, hardwood, etc.) and paper.[1/] This chapter examines historical growth trends in the demand for these products, analyzes the technological and other factors which have shaped demand development, and presents indicative demand projections for the year 2000. Principal conclusions developed in this chapter are:

(a) Demand for all primary forest products has increased over the past thirty years but the pattern of growth has differed from product to product. Sawnwood is becoming of decreasing importance in the world economy, while panels increased in importance through the 1950s and 1960s but for most types of panel, demand growth is now close to overall economic growth. Paper has maintained a more-or-less constant position in the world economy. In general, growth for all products is stronger in developing countries than in industrial countries.

(b) Median projections of world consumption of forest products are as follows:

| | Actual 1984-86 Average | Projected 2000 | Average Annual Growth |
|---|---|---|---|
| Sawnwood (million cu m) | 463 | 538 | 1.0% |
| Panels (million cu m) | 111 | 156 | 2.3% |
| Paper (million tons) | 195 | 299 | 2.9% |

These projections are somewhat less optimistic than most others which have been prepared in recent years, reflecting a revised assessment of the degree of maturity of markets in industrial countries.

### Global Trends

Although sawnwood, panels and paper are all based on wood, the three products are at significantly different stages of market development. In terms of its share in world-wide economic activity, the importance of sawnwood has halved in the past thirty years; panels have grown in importance (but now seem to have reached a plateau); while paper has maintained a relatively constant share. These trends are illustrated in Table 2.1 which shows the development since 1955 of world consumption of the three primary products per million dollars of world GNP.

---

1/ Annex A contains a short glossary of products and processes of the forest industry.

Table 2.1

World: Consumption of Forest Products per Unit of GNP
(units consumed per million 1980 US$)

| | 1955 | 1965 | 1975 | 1985 |
|---|---|---|---|---|
| Sawnwood (cubic meters) | 91 | 71 | 50 | 42 |
| Wood-based panels (cubic meters) | 6.1 | 8.9 | 10.7 | 10.2 |
| Paper/paperboard (tons) | 17 | 18 | 17 | 18 |

Sources: Consumption data are three-year averages from "Yearbook of Forest Products". FAO, Rome, various years. GNP estimates developed from "World Development Report 1986", Oxford University Press, New York, 1986.

Because of these diverse trends, representing different levels of maturity in the markets for each of the products, it is appropriate to separately examine historical demand trends and future prospects.

## Sawnwood

World consumption of sawnwood 2/ has increased over the past thirty years, and even over the past ten years, when demand for many other commodities has been declining. However, in recent years, the principal element of this overall growth trend has been the relatively vigorous growth in demand in developing countries (see Table 2.2).

While the slowing rates of demand growth in industrial countries are partly attributable to the slowdown in world economic growth, changing consumption patterns have had an even more marked effect, as evidenced by declining elasticities of demand with respect to income in these countries. This contrasts with the situation in developing countries where elasticities are still increasing (see Table 2.3).

Reference to specific country situations gives similar results. Recent studies of the forest industries sectors of Chile and Indonesia give income elasticities of demand for sawnwood of 0.8 for both countries based on consumption statistics over the past 20 years. Over this same period, the elasticity for the United States is found to be -0.2. For the Federal Republic of Germany, and Japan, the elasticity is close to zero (slightly negative for Germany, and slightly positive for Japan).

---

2/ No differentiation is made here between "softwood" (from coniferous species) and "hardwood" sawnwood. Although the end-uses are generally rather different - softwood is used primarily for construction, while hardwood is also used for furniture, panelling, etc. - the consumption trends are similar: declining in the industrial countries and growing healthily in developing countries.

Table 2.2

Development of World Demand for Sawnwood, 1955-85

| | Apparent Consumption (million cubic meters) | | | | Average Annual Increase ------(percent)------ | | |
|---|---|---|---|---|---|---|---|
| | 1955 | 1965 | 1975 | 1985 | 1955 -65 | 1965 -75 | 1975 -85 |
| Developing Countries a/ | 32 | 41 | 65 | 105 | 2.3 | 4.8 | 4.9 |
| Industrial Countries | 275 | 332 | 356 | 358 | 1.9 | 0.7 | 0.1 |
| World Total | 307 | 373 | 421 | 463 | 1.9 | 1.2 | 1.0 |

a/ FAO designations have been used in classifying "developing" and "industrial countries. Details are contained in Annex B.

Source: Annex C, Table C.1.

Table 2.3

Development of Income Elasticities of Demand for Sawnwood

| | Observed Elasticities | | |
|---|---|---|---|
| | 1955-65 | 1965-75 | 1975-85 |
| Developing Countries | 0.5 | 0.8 | 1.1 |
| Industrial Countries | 0.4 | 0.2 | 0.0 |
| World Total | 0.4 | 0.3 | 0.3 |

Source: Annex C, Table C.1.

An understanding of the underlying reasons behind these trends is necessary for an appreciation of the forces which have shaped the development of sawnwood markets, which, in turn, is essential if reasonable prognostications for the future are to be made. The principal use of sawnwood is construction, and particularly residential construction. In the industrial countries, slowing rates of population growth, and the development of an adequate stock of housing in the post-war years, have meant a slowing and stabilization of the rate of new house construction. In the United States, for example, new housing starts have generally cycled from 1 to 2 million units annually over the past twenty-five years, with the highest figure during this period being 2.4 million units reached in 1972.

A second evolutionary trend is the development of alternative materials to sawnwood for many uses. Most importantly, panels such as plywood and particleboard have replaced sawnwood for such uses as concrete forms, flooring underlay, and wall and roof sheathing in woodframe construction. The use of aluminum studs for building is another substitution which is eroding sawnwood markets, as is the use of particleboard (sometimes with a wood veneer) in furniture construction. With these types of development it is not surprising that sawnwood demand in industrial countries has been stationary or declining. Such trends are not yet evident in most developing countries where sawnwood is usually available at a considerably lower cost than alternative materials. For the forseeable future, sawnwood consumption in these countries is likely to proceed at least at the rate of economic growth.

## Wood-based Panels

Consumption of wood-based panels has increased steadily (at an average rate of 5.8% annually) over the past thirty years. As with sawnwood, demand growth has been particularly significant in developing countries but has also been sustained (albeit at declining rates) in industrial countries. The panel market is somewhat more complex than that for sawnwood because it includes veneer and plywood, which are produced directly from whole logs, and panels reconstituted from wood particles (particleboard, fiberboard). Because of some inherent differences in the development of markets for these two basic categories of board, the historical evolution of demand for each is shown separately in Table 2.4.

As with sawnwood, some of the decline in the rate of growth in demand can be attributed to slowing rates of growth of the world economy but in the industrial countries, there has also been a marked decline in the elasticity of demand for panels with respect to income as illustrated in Table 2.5. The figures in the table suggest some degree of saturation in demand for panels in industrial countries, although there is, as yet, no indication of this in developing countries.

### Plywood

Demand for plywood in industrial countries has developed partly as a result of overall economic (and particularly construction) activity, and partly as plywood has replaced sawnwood in many construction applications. However, as new opportunities for such substitution have dried up, the rate of growth in the demand for plywood, which was very healthy through the 1950s and 1960s, has slowed considerably. This situation has been exacerbated by the advent of particleboard, wafer board, and oriented structural board which, being cheaper products, have in turn replaced plywood in many of these applications. In the future, in

Table 2.4

Development of World Demand for Wood-based Panels, 1955-85

| | Apparent Consumption (million cubic meters) | | | | Average Annual Increase -----(percent)----- | | |
|---|---|---|---|---|---|---|---|
| | 1955 | 1965 | 1975 | 1985 | 1955 -65 | 1965 -75 | 1975 -85 |
| Developing Countries | | | | | | | |
| - plywood/veneer | 0.7 | 1.7 | 4.0 | 10.0 | 9.4 | 8.7 | 9.7 |
| - reconstituted panels | 0.1 | 0.7 | 2.6 | 5.6 | 22.9 | 13.8 | 8.1 |
| Industrial Countries | | | | | | | |
| - plywood/veneer | 11.8 | 24.7 | 36.8 | 39.0 | 7.7 | 4.1 | 0.6 |
| - reconstituted panels | 7.9 | 19.9 | 46.2 | 56.7 | 9.6 | 8.8 | 2.1 |
| World | | | | | | | |
| - plywood/veneer | 12.5 | 26.4 | 40.8 | 49.0 | 7.8 | 4.4 | 1.9 |
| - reconstituted panels | 8.0 | 20.6 | 48.7 | 62.2 | 9.9 | 9.0 | 2.5 |
| World Total | 20.5 | 47.0 | 89.5 | 111.2 | 8.7 | 6.7 | 2.2 |

Source: Annex C, Table C.2.

Table 2.5

Development of Income Elasticities of Demand for Wood-based Panels

| | Observed Elasticities | | |
|---|---|---|---|
| | 1955-65 | 1965-75 | 1975-85 |
| Developing Countries | 2.4 | 1.9 | 2.1 |
| Industrial Countries | 1.8 | 1.6 | 0.6 |
| World Total | 1.8 | 1.6 | 0.8 |

Source: Annex C, Table C.2.

industrial countries, it seems likely that the demand for plywood (like sawnwood) will not grow very rapidly, if at all, and may even continue (as in recent years) to decline.

No such trend is likely in the developing countries, at least in the medium-term future. Particularly in countries with a domestic plywood industry, rapid growth in domestic demand (such as the ten-fold increase in Indonesia over the past ten years) is likely.

Reconstituted wood panels

For this product group, further disaggregation is necessary because of the diverse characteristics of the three main products: particleboard, compressed fiberboard (hardboard), and non-compressed fiberboard.[3/] As illustrated in Table 2.6, almost all of the growth in this industry in recent years can be attributed to particleboard. Particleboard now accounts for 75% of industrial countries' demand for this type of product, compared with only 5% in 1955.

Table 2.6

Development of Industrial Countries' Demand for Reconstituted Wood-based Panels (1955-85)

| | Consumption (million cu m) | | | |
|---|---|---|---|---|
| | 1955 | 1965 | 1975 | 1985 |
| Particleboard | 0.4 | 8.6 | 30.8 | 42.5 |
| Compressed fiberboard | 1.6 | 3.9 | 7.0 | 7.4 |
| Non-compressed fiberboard | 5.8 | 7.4 | 8.4 | 6.7 |
| Totals | 7.8 | 19.9 | 46.2 | 56.6 |

Source: Annex C, Table C.2.

As with plywood, a major factor in the growth of demand for particleboard is the extent to which it has been a cheaper (and in some cases better) replacement for alternative materials, primarily sawnwood and plywood, and more recently (with the development of techniques for making thin particleboard), compressed fiberboard. Such substitutions have taken place both in construction and in more sophisticated end-uses such as the manufacture of furniture and cabinets. These latter uses in particular are likely to continue apace with world economic growth and so it is realistic to forecast that in industrial countries, there will be continued growth in the demand for particleboard.

---

3/ A third grade of board - medium density fiberboard (MDF) - has been attracting considerable interest of late for particular applications such as molded cabinets, picture framing, etc. Since 1980, capacity for this product has more than doubled, from 2.5 to 4.5 million cubic meters annually, of which about 15% is in developing countries. While still relatively small in relationship to particleboard, the rapid growth in the demand for this product, and its potential as a replacement for both wood and certain grades of particleboard, merits attention.

Compressed fiberboard is used for doorskins, furniture (particularly the backs of cabinets) and, with printed overlays, for low cost panelling. Replacement in some uses with thin particleboard, and growing consumer sophistication, could lead to erosion of markets for this product in the future although an optimistic scenario would see further modest growth. Non-compressed fiberboard is used for acoustic tiles and cheap wall-covering but demand has been declining steadily in industrial countries; this trend may be expected to continue.

In developing-country markets, where per capita consumption is so much lower, none of the above trends has yet been evidenced. Demand elasticities with respect to income have remained about 2: higher for plywood, and somewhat lower for fiberboard, particularly the non-compressed grade. Saturation levels are likely to be lower in most developing countries, partly because these panels may be less suitable for tropical countries in some applications, and partly because at least in the case of particleboard, a substantial part of the cost of production is chemical resins which normally have to be imported. However, for the medium term, with no indication (and no logical expectation) of saturation at present levels of consumption, it seems reasonable to assume that recent demand elasticity experience is appropriate for projecting future demand.

## Paper and Paperboard

As previously noted paper has maintained a relatively constant share of economic activity (about 18 tons per million US$ of GDP at 1980 prices) over the past 30 years. There are differences from grade to grade, and from country to country but, overall, world demand for paper has grown more or less apace with economic growth throughout this period. Table 2.7 shows how demand has developed by major grade grouping, in developing and industrial countries.

In the case of paper, the decline in the rate of growth of demand can, more-or-less, be directly attributed to slowing rates of growth in the world economy.

### Newsprint

In industrial countries, the income elasticity of demand for newsprint has remained at a fairly constant 0.8-0.9 over the past 30 years (Annex C, Table C.3). The principal determinant of newsprint consumption in the industrial countries is overall expenditure on advertising, and the proportion of this expenditure allocated to newspapers. In the United States, the early days of television saw some erosion of expenditures on newspaper advertising but since the 1960s, an almost constant proportion of around 28% of all advertising expenditures has been in newspapers, compared with around 20-22% on television. In Europe, although there are significant differences from country to country, typically one half of all advertising expenditures are for newspapers and magazines. Since advertising expenditures are rather closely tied to the overall level of economic activity, it is not surprising that income elasticity of demand

Table 2.7

Development of World Demand for Paper, 1955-85

| | Apparent Consumption (million tons) | | | | Average Annual Increase(%) | | |
|---|---|---|---|---|---|---|---|
| | 1955 | 1965 | 1975 | 1985 | 1955 -65 | 1965 -75 | 1975 -85 |
| Developing Countries | | | | | | | |
| - newsprint | 0.9 | 1.7 | 2.4 | 3.8 | 6.2 | 3.2 | 4.8 |
| - printing/writing | 0.7 | 1.9 | 4.1 | 8.3 | 10.9 | 8.3 | 7.2 |
| - other grades | 1.2 | 4.5 | 9.9 | 20.5 | 13.8 | 8.3 | 7.5 |
| Industrial Countries | | | | | | | |
| - newsprint | 10.3 | 14.9 | 20.1 | 25.1 | 3.8 | 3.0 | 2.2 |
| - printing/writing | 9.6 | 17.1 | 26.8 | 43.3 | 6.0 | 4.6 | 4.9 |
| - other grades | 33.4 | 55.9 | 79.1 | 93.5 | 5.3 | 3.5 | 1.7 |
| World | | | | | | | |
| - newsprint | 11.2 | 16.6 | 22.5 | 28.9 | 4.0 | 3.0 | 2.5 |
| - printing/writing | 10.3 | 19.0 | 30.9 | 51.6 | 6.4 | 5.0 | 5.2 |
| - other grades | 34.6 | 60.4 | 89.0 | 114.0 | 5.7 | 4.0 | 2.5 |
| World Total | 56.1 | 96.0 | 142.4 | 194.5 | 5.5 | 4.0 | 3.2 |

Source: Annex C, Table C.3.

for this grade has remained close to unity.4/ However, the use of other grades of paper (e.g., coated grades for advertising flyers), declining rates of population growth and somewhat reduced newspaper circulation (which has recently been observed in some European countries) all suggest that a somewhat lower demand elasticity may be appropriate for projecting future demand.

Advertising expenditures are of less significance in developing countries, which consume about 10% of total newsprint consumption, but rising literacy rates and the general growth of communications have kept demand growing in these countries at a rate somewhat faster than the overall rate of economic development. This situation is expected to continue, at least in the medium-term future.

4/ Another technological change which has affected the tonnage of newsprint consumed is the reduction in basis weight which most producers undertook during the 1970s. By reducing basis weight from 52 grams per square meter to an average of about 47 grams per square meter, an increase of 10% of the printable area per ton of newsprint has been obtained.

### Printing and writing paper

Although the "paperless office" was projected to be the natural outcome of the rapid introduction of electronic business machines, until now the exact opposite has been the case. World consumption of wood-free printing and writing paper has grown at about twice the rate of growth of other grades, primarily as a result of the proliferation of high-speed copying machines, computer printers, etc. These developments can most clearly be seen by looking at the development of the demand elasticity over time (see Table 2.8).

Table 2.8

Development of Income Elasticities of Demand for Printing and Writing Paper

| | Observed Elasticities | | |
|---|---|---|---|
| | 1955-65 | 1965-75 | 1975-85 |
| Developing Countries | 2.3 | 1.5 | 1.7 |
| Industrial Countries | 1.3 | 1.2 | 1.9 |
| World Total | 1.3 | 1.2 | 1.8 |

Source: Annex C, Table C.3.

Until 1975, the trend in elasticity is similar to that for other forest products: a world-wide downward trend, driven by the downward trend in industrial countries as saturation levels are approached. Then, from 1975, a turnaround to higher elasticities is evident. It is not realistic to assume that such rapid growth will continue into the foreseeable future. As more sophisticated networking of computer systems is introduced, the expected impact of electronic competition on office paper flows may be felt. There is little evidence of this as yet, however, and no general agreement among industry specialists as to when it might happen.

### Other grades

The rather large "other" category in Table 2.7 embraces a very wide range of paper types including household and sanitary (tissue) papers (about 10%), corrugated containerboard (40%), wrapping grades (20%), folding boxboard (15%), and other unspecified grades (15%). Although the market trends and prospects for these grades are rather different, international statistical information is inadequate to allow individual assessments of demand prospects. Overall, however, there has been a downward trend worldwide in demand elasticities (see Table 2.9).

Table 2.9

Development of Income Elasticities of Demand for Other Grades of Paper

| | Observed Elasticities | | |
|---|---|---|---|
| | 1955-65 | 1965-75 | 1975-85 |
| Developing Countries | 2.8 | 1.5 | 1.7 |
| Industrial Countries | 1.1 | 0.9 | 0.7 |
| World Total | 1.2 | 1.0 | 0.9 |

Source: Annex C, Table C.3.

Particularly in industrial countries, the demand for packing and wrapping grades, the largest component of this category, has been affected by a number of evolutionary factors. These include the loss of some markets to competing (especially plastic) packing materials, saturation of traditional markets in some countries, streamlining of distribution systems and inventory control which reduce packaging needs, government regulations which encourage the use of returnable packaging, simplification and elimination of packaging as a cost-saving measure, and the promotion of bulk transportation. In some developing countries reverse trends are noted, such as the replacement of cotton sacks with paper, and wooden boxes with containerboard.

The general consensus among industry specialists is that in industrial countries, demand for these grades will continue to grow, but at a rate significantly lower than overall economic growth. In developing countries, where consumption levels are so much lower, continued healthy growth is anticipated.

## Indicative Projections of Demand

To make any useful prognosis about the future of the world's forest industries, it is essential to have reasonable projections of the future demand for the major primary forest products. Projections for this paper have been prepared on the basis of historical consumption experience and evolving technological and market trends as discussed above, which have been projected together with estimates of future economic growth. Market trends have been characterized in terms of projected demand elasticities, with a high and low range for each product. For economic growth, indicative World Bank high and low projections have been assumed.[5/] For

5/ "World Development Report". Oxford University Press, New York, 1986 (page 45).

the low assumption, these are 2% per year for all countries. For the high assumption, 3.8% has been used for industrial countries, and 3.9% for developing countries. Although these projections cover only the period to 1995, they have been extrapolated to 2000.

Detailed results of the projections are contained in Annex C, Table C.4. Because of the considerable range, both in projected economic growth rates, and in elasticities, and the compounding effect of combining the assumptions for each factor, there is also a considerable range between the low and high projections for some products. For indicative planning purposes the median of the two projections is adequate and is summarized in Table 2.10.

Table 2.10

Projected Demand for Forest Products, 1995 and 2000

| | Developing Countries | | Industrial Countries | | World | | |
|---|---|---|---|---|---|---|---|
| | 1995 | 2000 | 1995 | 2000 | 1985 (Actual) | 1995 | 2000 |
| Sawnwood (million cu m) | | | | | | | |
| - hardwood | 83 | 99 | 55 | 53 | 116 | 138 | 152 |
| - softwood | 63 | 76 | 307 | 310 | 347 | 370 | 386 |
| Total sawnwood | 146 | 175 | 362 | 363 | 463 | 508 | 538 |
| Panels (million cu m) | | | | | | | |
| - plywood/veneer | 17 | 22 | 41 | 42 | 49 | 58 | 64 |
| - particleboard | 6 | 8 | 57 | 67 | 46 | 63 | 75 |
| - fiberboard | 3 | 4 | 13 | 12 | 16 | 16 | 16 |
| Total panels | 26 | 34 | 111 | 121 | 111 | 137 | 155 |
| Paper/paperboard (million tons) | | | | | | | |
| - newsprint | 5 | 6 | 31 | 35 | 29 | 36 | 41 |
| - printing/writing | 13 | 17 | 65 | 81 | 52 | 78 | 98 |
| - other grades | 33 | 43 | 109 | 118 | 114 | 142 | 161 |
| Total paper/paperboard | 51 | 66 | 205 | 234 | 195 | 257 | 300 |

Source: Annex C, Table C.4.

These projections suggest that the demand for sawnwood will grow at about 1.0% annually through the end of the century, wood-based panels at 2.3%, and paper at 2.9%. These are not very high expected rates of growth, and are lower than the expectations of some other forecasters as shown by the comparison in Table 2.11.

Table 2.11

Comparison of Projected Rates of Growth in Demand of Forest Products
(Average annual growth rate to 2000, percent)

| | Sawnwood | Panels | Paper |
|---|---|---|---|
| This study (median estimate) | 1.0 | 2.3 | 2.9 |
| FAO Projections | | | |
| FAO staff (1982) a/ | 1.2-1.6 | 4.4-5.2 | 3.7-4.0 |
| Industry Working Party (1982) a/ | 1.1 | 2.2 | 3.5 |
| FAO staff (1986) b/ | 1.6-2.0 | 4.1-5.0 | 2.6-3.1 |
| FAO staff (1986 - projection to 1995) c/ | -e/ | -e/ | 2.6-2.9 |
| Industry Working Party (1986 - projection to 1995) c/ | -e/ | -e/ | 2.3 |
| World Bank (1984--hardwood sawnwood only) d/ | 0.8 | -e/ | -e/ |

a/ "World Forest Products: Demand and Supply, 1995 and 2000". FAO Forestry Paper No. 29, FAO, Rome, 1982.
b/ "Forest Products: World Outlook Projections". FAO Forestry Paper No. 73, FAO, Rome, 1986.
c/ "The Outlook for Pulp and Paper to 1995", FAO, Rome, 1986.
d/ World Bank data.
e/ No estimate.

The generally lower projection reflects, first, the observation that demand elasticities for almost all products are falling, and are zero or negative in the industrial countries for both sawnwood and plywood, and, second, a conviction that this trend will not be reversed. It is of interest to note that, at least for panels and paper, FAO projections have tended to become lower with time. Also, the Industry Working Parties which FAO assembles to make subjective assessments of future market trends, tend to project lower growth than FAO itself. For the purposes of this paper, the median estimate developed above have been used to determine likely industry developments, such as pressures on forest resources, and capital investment requirements.

## Investment Implications

Before turning to the question of resource availability to meet this demand, and comparative economics to assess where additional capacity may be installed, it is interesting to translate these projected increases in demand into at least an indicative projection of capacity additions. First, for sawnwood, net additions to capacity will be considerably less than additional demand, because there is already substantial excess capacity in this industry. Assuming, however, that half of the increase in demand will be supplied from the output of new sawmills of an average of 50,000 cubic meters annual capacity, this implies the construction (world-wide) of about 50 new sawmills a year, which at a capital cost of perhaps US$20 million per mill (in 1987 prices), will require investments of US$1,000 million annually for sawmill construction in each of the next 15 years.

For panels, approximately 8 plywood mills of 100,000 cubic meters annual capacity and 9 particleboard mills of 200,000 cubic meters annual capacity will be required each year until the end of the century, implying investments of a further US$600 million annually.

For pulp and paper, which is more capital intensive and for which higher growth rates are expected, annual capacity additions could include 4 newsprint mills (of 180,000 tons annual capacity, and costing US$400 million each); 12 integrated pulp and paper mills (150,000 tons and US$500 million each); 5 market pulp mills (250,000 tons and US$500 million each); and about 40 small to medium-sized, non-integrated mills, which it has been assumed would average 40,000 tons annual capacity and cost an average of US$35 million each. These additions would meet about 80% of additional capacity requirements for a total cost of US$12 billion annually. It has been assumed that the other 20% would be met by expanding and improving existing mills, requiring annual investments of about US$1 billion.

In all, therefore, the world's forest industry will require expenditures for new capacity of about US$15 billion annually (1987 prices) through the end of the century to meet anticipated increases in demand. Subsequent chapters of this report will attempt to identify the factors which are likely to determine where these expenditures might be made.

# III. THE FOREST RESOURCE

There is about one half of a hectare of commercial forest for every inhabitant of the earth. The forests are not very equitably distributed, however: inhabitants of Canada have more than 12 ha per capita, while in Bangladesh there is only one hundredth of a hectare per person. This chapter reviews the extent and location of the world's forest resources and the rate at which they are being depleted or extended. The analysis concludes that while there are serious shortages of wood for fuel and building in many parts of the world, global forest resources are adequate to supply the needs of forest industries into the foreseeable future.

## The World's Forests

The world's natural "closed" 6/ forests cover about 2,800 million ha, or 22% of the world's land area. Out of this total, about 700 million ha is non-commercial forest, either because it is inaccessible or because it is reserved for other purposes such as a national park. In addition to natural forests, an estimated 48 million ha of industrial plantations have been established. Details on a country by country basis of commercial natural forest an plantations are summarized in Table 3.1.

Certain facts concerning these areas are noteworthy:

(a) In North America and Europe, although there has been massive destruction of forests in past centuries, the situation is now more or less stable, and forested areas are essentially constant or increasing slightly;

(b) The tropical forests in Asia, Africa and Latin America are being destroyed at a rate of about 8 million ha annually, primarily as a result of shifting cultivation, and land-clearing for agriculture; and

(c) New industrial plantations are being established at a rate of about 2 million ha annually.

## Current Demands on Forests

While forests are the source of basic raw material for forest industries, they are at the same time of vital ecological importance for catchment protection, wildlife conservation and erosion control, and a source of materials for many important non-industrial activities, in particular as fuel, but also for agriculture, building poles, weaving

---

6/ The term "closed" forests refers to closure of the crowns of the trees, and implies relatively full stocking as opposed to open woodland or savanna formations.

Table 3.1

World Commercial Forest Resources (1985)
(millions of hectares)

| Region | Natural Forests: Coniferous | Natural Forests: Broad-leaved | Natural Forests: Bamboo | Plantations: Coniferous | Plantations: Broad-leaved | Total Commercial Forests | Percent of Total |
|---|---|---|---|---|---|---|---|
| Africa | 1 | 156 | 1 | 2 | 1 | 161 | 7 |
| North America | 374 | 135 | - | - | - | 509 | 24 |
| Latin America | 15 | 550 | - | 4 | 2 | 571 | 26 |
| Asia | 69 | 221 | 9 | 18 | 17 | 334 | 16 |
| Europe | 78 | 56 | - | 1 | 2 | 137 | 6 |
| USSR | 320 | 70 | - | - | - | 390 | 18 |
| Oceania | - | 54 | - | 2 | - | 56 | 3 |
| World Totals | 857 | 1,242 | 10 | 27 | 22 | 2,158 | 100 |
| Of which: | | | | | | | |
| - Developing | 73 | 929 | 10 | 24 | 19 | 1,051 | 49 |
| - Industrial | 784 | 313 | - | 3 | 3 | 1,107 | 51 |

Source: Annex C, Tables C.5-C.8. The reliability of the estimate, and also the definition, in terms of the distinction between natural and plantation forest, varies from country to country.

materials, sericulture, etc. Most types of utilization of forests for industrial purposes are not in conflict with these other activities, but uncontrolled exploitation is of major concern, particularly in tropical rain forests.

Estimates of world-wide consumption of wood, both for fuel and for industry, are related to regional forest areas in Table 3.2. It should be noted that because non-commercial forest areas are not shown, and at least some of the fuelwood removals are from such areas, the removal rate for fuelwood in cu m/hectare is somewhat overstated.

Some interesting deductions can be made from the statistics in Table 3.2.

(a) Fuelwood accounts for somewhat more than one half of total removals;

(b) The intensity of harvesting for industrial wood is highest in Europe (2.1 cu m/hectare) and North America. Extraction in these regions is not, in most cases, placing undue pressure on the forests; and

(c) The overall rate of extraction (about 3,000 million cubic meters annually) is significantly less than even the most cautious estimate of annual growth, which for commercial forests alone would be at least 4,000 million cubic meters annually.

Table 3.2

Regional Forest Areas and Removals (1985)

| Region | Commercial Forest Area (million ha) | Removals: Fuel (million cu m) | Removals: Industry (million cu m) | Removal Rate: Fuel (cu m/hectare) | Removal Rate: Industry (cu m/hectare) | Removal Rate: Total (cu m/hectare) |
|---|---|---|---|---|---|---|
| Africa | 161 | 399 | 54 | 2.5 | 0.3 | 2.8 |
| North America | 410 | 108 | 512 | 0.3 | 1.2 | 1.5 |
| Latin America | 559 | 265 | 94 | 0.5 | 0.2 | 0.6 |
| Asia | 337 | 734 | 244 | 2.2 | 0.7 | 2.9 |
| Oceania | 58 | 8 | 28 | 0.1 | 0.5 | 0.6 |
| Europe | 127 | 56 | 293 | 0.4 | 2.3 | 2.8 |
| USSR | 535 | 80 | 275 | 0.1 | 0.5 | 0.7 |
| Totals | 2,186 | 1,650 | 1,500 | 0.8 | 0.7 | 1.4 |

Source: Annex C, Tables C.5-C.8.

Despite this apparently satisfactory perspective on a world-wide basis, the wood-supply situation is critical in certain countries and geographic regions. As detailed in Annex C, Table C.5, in twenty-seven developing countries (twenty-two of which are in Africa) annual extraction of all types of wood exceeds 10 cubic meters per hectare of closed forest. These countries will make no contribution to future requirements of industrial timber, but more importantly, the lives of their people are disrupted by floods, fuelwood scarcity, soil and water degradation and reduced agricultural productivity, all of which are at least partly as a result of deforestation. While the extraction of timber for industrial uses is not a primary factor in this degradation, it can have a contributory effect where, for example, logging for commercial species is undertaken without adequate care to protect the timber which remains; or roads built to open up new areas for logging facilitate the migration of shifting cultivators; or where improperly designed roads and work areas contribute to erosion. These issues are examined in more detail in Chapter V of this paper.

Projected Future Demands for Industrial Wood

On the basis of the projections of demand for sawnwood, panels and paper presented in Chapter II, projections of roundwood requirements have also been prepared. The detailed projections are in Annex C, Table C.9 and summarized In Table 3.3.

Table 3.3

Projected Future Industrial Roundwood Requirement
(million cubic meters)

| | Actual 1985 | Projected 2000 | | |
|---|---|---|---|---|
| | | Low | High | Median |
| Sawlogs & veneer logs | 898 | 963 | 1,153 | 1,058 |
| Pulplogs | 387 | 512 | 811 | 660 |
| Other industrial wood a/ | 235 | 230 | 230 | 230 |
| Total | 1,520 | 1,705 | 2,194 | 1,948 |

a/ The consumption of "other industrial wood" has been assumed to be constant, reflecting the trend of recent years.

Source: Annex C. Table C.9.

Assuming the median projection, the additional requirement for industrial roundwood will be some 428 million cubic meters annually by 2000, an increase of about 28% over present supply.

Supply Potential

The actual wood supply situation at the end of the century will depend on more than mere availability of wood, and various alternative supply scenarios are examined in Chapter VI of this report. However, in order to prepare such scenarios, it is necessary to know the potential supply from various countries and regions. The following assessments are based largely on the work of FAO 7/ although when FAO attributes other sources, these are noted.

7/ "World Forest Products: Demand and Supply, 1990 and 2000." FAO, Rome, 1982.

Industrial countries

In industrial countries, productivity gains, and new areas, have the potential to increase industrial log production at the end of the century by about 390 million cubic meters annually. For the main producing countries and regions, the basis of these assessments is summarized in the following paragraphs.

(a) According to Forest Service projections, the United States' forests have the potential to sustain an annual output at least 100 million cubic meters more than at present.

(b) By opening up new areas, improving forest management and increasing the use of forest residues, annual log output from Canada could increase by about 40 million cubic meters by the end of the century.

(c) According to the ECE/FAO Agricultural and Timber Division, the forested area of Western Europe will increase by about 4% (6 million ha) by the end of the century. By the end of the century, this increase, coupled with improved productivity, will allow removals to increase by about 60 million cubic meters annually.

(d) Japan has undertaken major reforestation programs throughout the country since the end of the war, mostly softwoods on private land. Industrial log output is expected to increase by about 20 million cubic meters annually by 2000.

(e) According to FAO, Eastern Europe's production of industrial roundwood will increase by up to 15 million cubic meters annually by the end of the century. Over the same period, an additional 120 million cubic meters annually could be supplied by the Soviet Union.

(f) Australia, New Zealand and South Africa all have significant and expanding plantations which supply pulplogs and sawlogs for export as well as domestic consumption. The incremental output from these plantations by 2000 is expected to be about 35 million cubic meters annually.

Developing countries

3.11 The potential in developing countries is not as great, and in some cases, particularly in natural tropical forests, the estimates of "Annual Allowable Cut" are based on assumptions concerning regrowth which may not be supportable. Distinction must be made here between industrial wood from natural tropical forests, and the rapidly expanding production of pulplogs (and to a lesser extent sawlogs and veneer logs) from industrial plantations established in many countries over the past 20-30 years. The potential for expanding sawlog and veneer log production from tropical forests is estimated to be some 20 million cubic meters annually by the end of the century. Production from plantations could grow by 165 million cubic meters annually over the same period. Notes on the potential in the main producing areas and countries are provided in the following paragraphs.

Considering tropical hardwood logs, the traditional African suppliers in Ivory Coast, Ghana and Nigeria have reached the limit of their production potential because of past over-cutting. On the other hand, Liberia, Cameroon, Gabon and Congo have potential to increase production as, in the longer term, do Angola, Mozambique, Equatorial Guinea and Zaire. In Latin America, Amazonia has tremendous potential, but at rather higher costs. In Asia, Sarawak and Sabah may be approaching their potential output by the end of the century but Indonesia's reserves are largely untouched. A recent ban on log exports in Indonesia (to promote domestic processing) should facilitate preservation of this resource. It is projected that by the end of the century, the annual output of tropical hardwood logs will increase by about 4 million cubic meters from Africa, 5 million cubic meters from Asia, and 12 million cubic meters from Latin America, for a total increment of about 20 million cubic meters.

Estimates of industrial wood production from plantations is a little more complex. The major developing country producers are in Latin America, where Brazil and Chile alone will have the potential to increase industrial log production by about 75 million cubic meters annually by the end of the century. Other Latin American countries could add another 25 million cubic meters to this total. With the exception of China and India, which have very large areas of low-productivity plantations, industrial tree planting is not nearly as widespread in Asia as in Latin America. In Africa (apart from South Africa) there are reported to be about 1.8 million ha of industrial plantations, more or less equally split between hardwoods and softwoods. If these plantations are maintained on a sustained yield basis, assuming an average mean annual increment (MAI) of 12 cubic meters per hectare, they should be capable of producing about 20 million cubic meter annually by 2000, up from perhaps 5 million cubic meters annually today. Similar calculations in Asia require the use of a significantly lower MAI, but still suggest that industrial wood production from plantations in this area could increase by around 50 million cubic meters by the end of the century.

## Development Implications

From the foregoing, it can be seen that while demand for industrial timber will increase by an estimated 428 million cubic meters annually by 2000, the supply potential is 390 million cubic meters in the industrial countries, and 185 million cubic meters in the developing countries. It is unlikely that there will be any overall shortage of industrial wood, which suggests first, that any country which intends to create new forest resources to compete in world markets had better be sure that it will be competitive, and second, that even to meet domestic demand, the establishment of new forests may not be economic in comparison to purchasing wood or wood products from other efficient producers. The question of comparative advantage and policy implications are examined in the following chapters.

# IV. GLOBAL COMPETITIVENESS AND TRADE

Wood is the common factor in forest products and for most products, in most situations, wood is the single most important element in the cost of production. As noted earlier, most developing countries can produce wood more quickly and more cheaply than most industrial countries. However this natural comparative advantage is not of itself sufficient to ensure that developing countries with cheap wood can necessarily be efficient producers of forest products, or even when they are, that this will ensure access to world markets.

This chapter presents a two-approach analysis of comparative advantage. First, trade patterns are examined. By examining historical trade flows and prices, and discussing the factors which have affected trends in these patterns, an historical perspective of comparative advantage is developed which provides valuable insight for future prospects. Second, a geographical comparison of the structure of production costs is presented. The principal conclusions which are derived in this chapter are that: (i) if trade is any indicator, comparative advantage in world markets belongs to the industrial countries where 86% of all exports of forest products originate; (ii) many developing countries can produce sawnwood from plantation wood more cheaply than most industrial countries, but differential selling prices erode or eliminate this apparent advantage; (iii) these same countries can produce pulpwood and wood-pulp more cheaply, and this advantage persists into the final product; and (iv) some grades of paper may also be made more cheaply in developing countries, but the margin may not be enough to warrant the marketing difficulties.

## Historical Development of World Trade

Trade in forest products has been expanding more quickly than production and consumption. If comparative advantage determines trading patterns, then an historical analysis of trading patterns will provide at least some insight into historical comparative advantage. This is particularly important: (i) where geographical considerations provide particular advantages, such as trade within Europe, and between Canada and the US; (ii) where ownership influences trade, such as US ownership of much of Canada's forest products industry, or Japanese investments in forest products companies in Canada, Brazil, New Zealand and elsewhere; and (iii) where marketing agreements (the EEC, COMECON), or an individual country's restrictive trade practices inhibit free trade. Under circumstances such as these, comparative advantage in terms of cost does not assure market share.

Some basic statistics on trade in forest products in 154 countries are presented in Annex C, Tables C.10-C.13. Regional data are summarized in Table 4.1.

Table 4.1

The Structure of Trade in Forest Products (1985)

| Region | Trade in Forest Products -----(US$ billions)----- Imports | Exports | Net | Forest Products in Total Trade ----(percent)---- Imports | Exports |
|---|---|---|---|---|---|
| Africa | 2.0 | 1.0 | (1.0) | 2.6 | 1.1 |
| North America | 11.6 | 16.6 | 5.0 | 2.6 | 5.5 |
| Latin America | 1.9 | 1.4 | (0.5) | 2.6 | 1.4 |
| Asia | 12.5 | 5.2 | (7.3) | 3.0 | 1.2 |
| Europe | 25.4 | 22.5 | (2.9) | 2.9 | 2.6 |
| Oceania | 1.0 | 0.6 | (0.4) | 3.0 | 2.0 |
| USSR | 0.9 | 2.7 | 1.8 | 1.1 | 3.1 |
| Totals | 55.3 | 50.0 | - | 2.8 | 2.6 |

Source: Annex C. Tables C.10-C.13.

Overall, forest products trade represents about 3% of world trade. Statistics for some individual countries are of interest. Among the industrial countries, fully 34% of Finland's export revenues derive from exports of forest products, while this figure is 16% in Sweden, and 13% in Canada. Among developing countries, Burma derives 28% of its export earnings from forest products (teak logs and sawnwood). Other major exporters are Paraguay (sawnwood and veneer--30% of total export earnings); Malaysia (logs, sawnwood and plywood--11%); Papua New Guinea (mainly logs - 9%); and Chile (all forest products--9%). While developing countries as a group show a modest surplus in trade in logs and mechanical wood products (about US$1 billion in 1985), this is more than offset by net imports of about US$4 billion for pulp and paper.

The following paragraphs discuss important aspects of trade in the main forest products groups.

### Logs and wood chips

While not itself a product of the forest industry, world trade in logs and wood chips is clearly of importance in any overall analysis of sectoral trends and prospects. Out of world production of around 1,500 million cubic meters of industrial wood, only some 100 million cubic meters is traded internationally in an unprocessed (or as wood chips) state. An approximate breakdown of this trade is shown in Table 4.2.

Table 4.2

World Trade in Unprocessed Wood (Average 1984-86)
(million cubic meters)

| Exporting Country/Region | Saw/Veneer Logs | | Other Indust. Wood | | Total |
|---|---|---|---|---|---|
| | Softwood | Hardwood | Softwood | Hardwood | |
| North America | 19 | 1 | 6 | 1 | 27 |
| USSR | 8 | - | 9 | - | 17 |
| Africa | - | 4 | - | - | 4 |
| Asia | - | 22 | - | 1 | 23 |
| Latin America | 1 | - | - | - | 1 |
| Europe | 4 | 3 | 9 | 8 | 24 |
| Oceania | 1 | - | 1 | 7 | 9 |
| Total | 33 | 30 | 25 | 17 | 105 |

Source: Derived from "Yearbook of Forest Products". FAO, Rome, 1987.

The biggest single trade flow in this category is the trade in tropical logs, with exports from Asia (Southeast Asia), primarily to other Asian countries (Japan, China, and Korea), and from Africa (West Africa), primarily to Europe. This trade is declining, partly because log supplies are being depleted (in Nigeria and the Philippines, for example), and partly because of restrictions on log exports which have been introduced by a number of countries (Malaysia, Indonesia), in order to protect or promote domestic processing. Those countries with tropical forests have a clear comparative advantage when it comes to trade in tropical logs, but the sensible decision in recent years to forgo the fast benefits of easy log exports for the more rewarding (but more difficult) alternative of establishing domestic industries should lead to a continuing decline of this trade.[8/]

8/ Indonesia, for example, has been very successful in this respect. In the late 1970s, Indonesia was the world's largest exporter of hardwood logs, mostly to plywood mills in Japan, China and Korea, earning close to US$1 billion annually in this trade. At that time, Indonesian plywood production was negligible. Restrictions on log exports were first applied in 1981, and since that time a domestic plywood industry has developed to the point that the country is now the world's largest exporter of plywood. Plywood exports in 1985 earned close to US$800 million.

A second major flow of unprocessed wood is the export of coniferous sawlogs from North America and the USSR, primarily to Japan, and to a lesser extent, to China and the Republic of Korea. New Zealand and Chile have also become suppliers in this market, exporting logs from their substantial plantations of fast-growing Pinus radiata. A comparison of prices paid in Japan for logs in this category is given in Table 4.3 which shows that in this market, not all suppliers are judged equal.

Table 4.3

Japan: CIF Prices for Coniferous Sawlogs (1985)
(US$ per cubic meter)

| Source and Log Type | CIF Price |
|---|---|
| Sitka spruce | |
| - USA and Canada | 141 |
| Douglas Fir | |
| - USA and Canada | 119 |
| Pine | |
| - USA and Canada | 115 |
| - Soviet Union | 83 |
| - New Zealand (Pinus radiata) | 75 |
| - Chile (Pinus radiata) | 58 |

Source: Japanese trade statistics.

The Japanese market clearly differentiates among the larger, slow-growing trees of North America, the rather scrubby Russian pine, and the fast-growing plantation pine of New Zealand and Chile. Although the plantation wood is much cheaper to produce, the advantage is lost in the price differential.[9/]

Other trade flows in Table 4.2 include primarily pulpwood to Japan from Australia, the USA and the USSR, and pulpwood within Europe. This is a growing market with relatively little differentiation in prices for different qualities of wood, and thus offers considerable scope for low-cost producers of wood. The principal elements of comparative advantage in pulpwood production are simply the cost of wood and the cost of freight.

---

9/ As will be further elaborated below, this price differential is worst for logs, but becomes of decreasing importance the more value is added. For paper, the New Zealand or Chilean product sells at essentially the same price as the North American competition.

Sawnwood

As noted in Chapter II, sawnwood consumption has been declining in industrial countries, although healthy growth in developing countries has contributed to a modest increase in world demand. Trade has been increasing and is now around 18% of total production, compared with only 11% in the mid-1950s. Total world trade in sawnwood is of the order of 80 million cubic meters annually, but about 30 million cubic meters of this total is between Canada and the USA,[10/] and a further 20 million represents trade amongst the countries of Europe, leaving only about 30 million cubic meters of extra-regional trade.

For softwood sawnwood, Western Europe, Japan and the Middle East are the main importing regions with North America and the USSR the principal exporters. The only significant developing-country producers in this market have been Honduras (with a natural Honduran pine), Brazil (natural Araucaria), and Chile (plantation pine). The natural stands of Honduras and Brazil are being depleted, but Chile's experience shows that there are prospects in world markets for low cost producers of sawnwood from plantation grown softwoods.

For hardwood sawnwood, most industrial countries are importers of tropical grades, principally from Ivory Coast, Cameroon and Ghana in West Africa, and Malaysia and Indonesia in Southeast Asia. For producers of sawnwood from tropical hardwood, as with logs, the natural comparative advantage lies with those who have the forests. In Africa in particular, there remains considerable scope to exploit further this advantage by increasing the level of domestic processing: of a dozen countries exporting tropical woods in Africa, only Ghana derives more export revenues from processed products than from logs. As previously noted, there are prospects for new producers to enter these markets but primarily because of more difficult access, costs would be higher.

Wood-based panels

World markets for wood-based panels differ somewhat from product to product. As can be seen in Table 4.4, developing countries are rather significant in the production and trade of veneer and plywood, and less so in the reconstituted panels.

---

10/ US sawnwood producers have long maintained that Canadian producers have been the beneficiaries of unfair comparative advantage in sawnwood production in that they were paying unreasonably low values for standing timber. Late in 1986, the USA agreed to impose a countervailing duty of 30% on all imports of Canadian sawnwood. Subsequently, further agreement has been reached with the Canadian Government that it would collect and retain the tax, and not use it to subsidize the sawnwood industry in any other way. This will effectively raise the domestic price of US sawnwood, and possibly lead to a reduction in Canadian export to the US.

Table 4.4

Wood-based Panels: Share of Developing Countries in Production and Trade (1985)
(percent)

| Panel Type | Developing Countries' Share | | |
| --- | --- | --- | --- |
| | Production | Imports | Exports |
| Veneer | 46 | 16 | 53 |
| Plywood | 28 | 33 | 70 |
| Particleboard | 7 | 4 | 2 |
| Compressed fiberboard | 20 | 8 | 20 |
| Non-compressed fiberboard | 4 | 10 | 0 |
| All panels | 18 | 18 | 39 |

Source: Annex C, Table C.2.

In the case of veneer, Malaysia alone accounts for one-half of all developing-country exports, with Congo and Paraguay being the only other significant suppliers. All of these countries are producing veneer from tropical hardwoods and, as with logs and sawnwood, have the natural comparative advantage of suitable forests for these markets. For plywood, there is a similar concentration with Indonesia accounting for around 60% of all developing-country exports, followed by China and Singapore (with 11% each), and the Philippines and the Republic of Korea (about 5% each). This market is interesting in that China, Singapore and Korea, without the natural comparative advantage of home-grown log supplies, managed to build up thriving plywood export businesses based on imported logs, primarily from Indonesia and Malaysia. In 1979, these three countries accounted for more than 70% of plywood exports from Southeast Asia. Now, due to the restriction of log exports, they supply only about 20%. In this case the comparative advantage was entrepreneurial spirit with a skilled, motivated, relatively low-cost work force. However, the advantage is lost when the log supply is curtailed and, in Korea in particular, many of the plywood mills which were built in the early 1970s, have been forced to close.

Amongst the reconstituted panels, developing countries are not a major factor in particleboard or non-compressed fiberboard, either in production or consumption. As previously discussed, this is partly because they are less suitable for some applications in developing countries, and partly because the cost of wood is a relatively small component in total production cost, so that cheap wood does not confer much of a comparative advantage. Compressed fiberboard is more widely produced and used, and Brazil in particular has been rather successful in exporting moderate amounts of this product to the USA where it has displaced domestic production.

Wood pulp

Almost 95% of paper is made from wood-pulp.[11/] Most commonly, the pulp is merely an intermediate material in an integrated pulp and paper mill but about 20% of all pulp is shipped from non-integrated pulp mills to non-integrated paper mills. In 1985, exports of wood pulp amounted to about 21 million tons, a figure which has been rising by 2-3% annually over the past decade. Almost 6 million tons out of this total is traded within Europe, and a further 4 million tons is exports from Canada to the USA, so that only about 11 million tons of wood pulp is traded extra-regionally. Half of this trade is shipments from North America to Europe and Japan, while the balance is exports to a wide variety of countries from the USSR (about 1 million tons annually), Brazil (1 million tons), Chile (0.5 million tons), New Zealand (0.5 million tons) and others. Total exports from developing countries amount to about 1.8 million tons (9% of all wood pulp exports) of which Chile and Brazil together account for more than 80%. The only other developing-country exporters of wood pulp are Swaziland, Morocco and China.

Wood pulp is an attractive commodity to consider for many developing countries. It can be produced from a wide range of woods, many of which grow rapidly in industrial plantations. It is a more-or-less standardized commodity, with many customers, and the technology is not particularly complicated. Unlike lower-valued mechanical wood products, freight can represent a fairly small part of the final delivered cost. On the other hand it is not possible to produce wood pulp competitively on a small scale, and capital costs can range into the hundreds of millions of dollars. Moreover, prices were rather depressed in the mid-1980s, as a result of new capacity coming on stream faster than the rate of demand increase. The cheap wood and large scale of operations in Brazil and Chile have led to profitable operations in these countries, but this is not necessarily a recipe that any country can follow.[12/] The cost comparisons developed later in this chapter under "Comparative Production Costs", show the extent of the comparative advantage which can accrue to a low-cost, wood-producing country, but this is not necessarily enough to ensure the success of such a venture.

---

11/ The remaining 5% is made from a variety of other fibers, mostly agricultural residues such as sugar-cane bagasse, and rice or wheat straw. About 30% of the fiber used for paper production is recycled waste but even this material originally came from wood.

12/ Malawi, for example, has an excellent plantation of pine pulpwood, and has tried to finance an export-oriented pulp mill for years. The extra cost of shipping pulp out of a land-locked country, and the huge investment required in relation to Malawi's economy proved to be insurmountable barriers. In another example, Cameroon did manage to finance such a mill but the company was under-capitalized, the mill was too small and the wood supply both expensive and technically unproven. The plant ran for a few months in 1983 but has subsequently closed down.

Paper

Trade statistics for paper yield only limited information about comparative advantage because on the whole, of the many grades of paper produced, relatively few are consistently traded internationally on a large scale. Data for the important categories are shown in Table 4.5.

Table 4.5

World Production and Trade of Paper and Paperboard (1985)
(millions of tons)

| Grade | World Production | World Trade: Within Europe | World Trade: Between Canada & US | World Trade: Other Trade | Total Trade |
|---|---|---|---|---|---|
| Newsprint | 28.3 | 3.9 | 7.6 | 2.3 | 13.8 |
| Printing/Writing | 50.9 | 5.6 | 1.1 | 3.0 | 9.7 |
| Tissue | 10.7 | 0.3 | - | 0.1 | 0.4 |
| Wrapping/Packaging | 75.8 | 7.3 | 0.5 | 2.8 | 10.6 |
| Other | 27.1 | 2.5 | 0.2 | 2.8 | 5.5 |
| Total | 192.8 | 19.6 | 9.4 | 11.0 | 40.0 |

Source: "Yearbook of Forest Products". FAO, Rome, 1987.

When the tightly organized trade within Europe, and between Canada and the USA (mostly newsprint exports from Canada) is excluded, only about 6% of world paper production is traded internationally. This is a clear indication that insofar as paper production is concerned, the principal comparative advantage arises from the existence of a domestic market. Important exceptions to this general observation are the markets for kraft paper and board, and for newsprint. The key to profitable production of these grades is low wood costs, large scale production, and preferably, minimal transport distances for finished products.

For kraft grades, growth in demand has been minimal in recent years and the present suppliers, principally in the US South, have so much capacity and are so efficient that new capacity for export in other locations can seldom be justified. For newsprint, Chile is the only developing country which has been able to sustain exports on any scale. Newsprint is also produced, but only for domestic consumption, in a number of other developing countries including Argentina, Brazil, Mexico, China, India, Bangladesh, the Republic of Korea and Turkey.

## Comparative Production Costs

Preceding sections have used trade statistics to demonstrate which producers appear to have real comparative advantage in world markets which can be translated into market share. This section will look more closely at actual costs and cost comparisons to see first, whether they adequately explain existing market structures, and second, whether they can be useful predictors of who could be competitive in future markets.

Costs are developed for a number of important forest products (sawnwood, particleboard, bleached chemical pulp and newsprint) in a number of producing countries. Two cautions are necessary in interpreting the data: first, costs are estimated averages, and may differ considerably from enterprise to enterprise within the specified country; and second, the comparisons are based on 1985 costs converted to US$ at 1985 exchange rates. Recent exchange rate shifts have had a considerable effect on comparative advantage, and this is examined as a separate exercise.

### Wood costs

The elements of wood cost are: (i) a "stumpage" or "royalty," representing the value of the standing tree at the time of felling; (ii) a harvesting cost; and (iii) a transport cost to the point of use. Approaches to establishing a stumpage value vary widely, and may depend on such factors as how the forest came into being (a natural or a man-made forest), the ownership, the cost of harvesting and transport, the value of the end-product, and the extent of any competition for the timber. Another approach, which is used most often by owners of forests to determine a transfer price to their own manufacturing facility, is to accumulate all of the establishment and maintenance costs, including a notional interest or rate of return, and to convert this to a cost per unit volume of wood obtained at the time of harvesting. This approach is also useful for comparing the cost of growing wood from one location to another, at different establishment costs and growth rates, and also for evaluating industrial plantation projects by determining a rate of return. Harvesting costs vary with the type and size of tree, the terrain and the technology employed. Transport costs vary with terrain, distance and equipment.

Table 4.6 gives examples of some recent wood costs figures for a variety of wood types delivered to a variety of locations. Figures in the table show that Chile and Brazil can produce the cheapest wood, due to a combination of abundant, relatively flat, relatively fertile land, within easy distance of possible sites for processing plants or ports. Both countries are already significant exporters of forest products, and are thus clearly taking advantage of this situation. However, the differences become less when full production costs, freight, and in some cases selling price, are taken into account.

### Sawnwood

For sawnwood, a simple two-country comparison is made, between softwood sawmills in Canada and Chile, shipping pine sawnwood to Japan (see

Table 4.6

Comparative Wood Costs (1985)
(US$ per cubic meter)

| Location & Wood Type | Stumpage | Harvesting | Transport | Total |
|---|---|---|---|---|
| Softwood Sawlogs | | | | |
| - Northwest USA | 14 | 12 | 14 | 40 |
| - Chile | 5 | 5 | 4 | 14 |
| - New Zealand | 14 | 7 | 5 | 26 |
| Hardwood Sawlogs | | | | |
| - Ivory Coast | 10 | 25 | 10 | 45 |
| - Congo | 9 | 25 | 52 | 86 |
| - Indonesia | 11 | 29 | 12 | 52 |
| Pulpwood | | | | |
| - Chile | 4 | 5 | 3 | 12 |
| - Brazil | 5 | 6 | 4 | 15 |
| - Southern USA | 13 | 8 | 6 | 27 |
| - Sweden | 20 | 10 | 10 | 40 |

Table 4.7). The products are not strictly comparable: Chilean sawnwood is made from smaller trees, and usually contains many knots. For some end-uses, the Chilean product would not be suitable but there is substitution at the margin, and the Japanese prices reflect the quality difference.

Although Table 4.7 shows the cost of sawnwood at the mill gate is considerably lower in Chile, the lower price paid because of the quality differential effectively eliminates any comparative advantage.[13/]

Particleboard

For particleboard, a notional comparison has been made between a low-cost wood country (say Brazil), and a high-cost but better-located country (Canada) shipping particleboard to the USA, which is a significant importer of this product (at present from Canada, but not from Brazil). The comparison is shown in Table 4.8.

---

13/ The quality differential is not necessarily permanent. Chilean forest owners have recently begun to introduce silvicultural and forest management practices (thinning, pruning, longer rotation) which will vastly improve the quality of sawlogs and hence sawnwood. The costs will be higher--perhaps US$5 per cubic meter of logs--but this would be adequately recompensed by a reduction in the quality differential.

Table 4.7

Comparison of Sawnwood Costs Delivered to Japan (1985)
(US$ per cubic meter)

| Cost Element | Canada | Chile |
|---|---|---|
| Log cost | 73 | 30 |
| (Net back for chip sales) | (16) | (9) |
| Energy | 2 | 3 |
| Labor & overheads | 20 | 21 |
| Other materials | 9 | 11 |
| Capital charges | 10 | 12 |
| Total at mill gate | 98 | 68 |
| Freight to Japan | 33 | 35 |
| Quality differential | - | 30 |
| Delivered Cost | 131 | 133 |

Table 4.8

Comparison of Particleboard Costs Delivered to the USA (1985)
(US$ per ton)

| Cost Element | Canada | Brazil |
|---|---|---|
| Fiber cost | 58 | 22 |
| Energy | 5 | 10 |
| Labor & Overheads | 33 | 21 |
| Chemicals & other materials | 77 | 89 |
| Capital charges | 15 | 15 |
| Freight | 19 | 34 |
| Delivered Cost | 197 | 191 |

Although wood in Brazil costs substantially less than in Canada, and although no price differential for quality reasons need be considered, for this product the comparative advantage for the low-cost wood producer represents only about 3% of the total product cost.

Pulp

For pulp, a somewhat more elaborate comparison has been made, including regional differences in Canada and the USA and Scandinavian producers as well as producers from New Zealand, Chile and Brazil. The detailed comparison is presented in Annex C, Table C.14, and for selected countries, summarized in Table 4.9.

Table 4.9

Comparison of Average Bleached Kraft Pulp Production Costs (1985) a/
(US$ per ton delivered to Western Europe)

| Cost Element | Chile | Sweden | Eastern Canada | Finland | USA South |
|---|---|---|---|---|---|
| Wood | 61 | 189 | 142 | 232 | 137 |
| Energy | 26 | 10 | 38 | 9 | 45 |
| Chemicals & other materials | 78 | 54 | 82 | 65 | 85 |
| Labor & Overheads | 39 | 54 | 83 | 63 | 83 |
| Capital charges b/ | 109 | 49 | 67 | 61 | 68 |
| Transport | 55 | 25 | 45 | 30 | 60 |
| Total | 368 | 381 | 457 | 460 | 478 |

a/ 1985 data compared at 1985 exchange rates.
b/ Capital charges include interest and depreciation. Chile's figures are higher because the capacity is relatively new.

Source: Annex C, Table C.14.

This comparison shows that a developing country with cheap wood such as Chile, despite cost disadvantages in both capital and transport charges, can deliver pulp more cheaply into Europe than any of these competitors. Moreover, this advantage primarily derives not from the effect of exchange rates, or subsidies, or any other such transitory effect, but from real climatic and topographic advantages.

Exchange rate variations can have a significant impact on these comparisons.[14/] Finland and Canada both ship several hundred thousand tons of pulp each year to West Germany. According to the mid-1985 comparison tabulated in Table 4.8, they are very closely competitive in terms of their total cost to deliver pulp to this market. However, if the analysis had been done in DM with the exchange rates which prevailed at the beginning of 1985, the Finns would have been shown to be more efficient producers, with costs about 12% lower than the Canadians. If the analysis were to be done today (January 1987), the Canadian costs (in DM) are a staggering 30% lower than the Finnish costs. In another example, during the late 1970s when Chile's exchange rate was seriously over-valued, its export-oriented pulp and paper industry got into very serious financial difficulties, from which it is only now recovering. A comparison using Chile's exchange rate at that time would not have shown Chilean producers to be nearly so competitive. Chile's exchange rate (like Canada's) is now more or less linked to the US dollar, and so the relative competitiveness of these suppliers is unlikely to change.

## Newsprint

A similar comparison for newsprint is detailed in Annex C, Table C.15, and summarized in Table 4.10.

### Table 4.10

Comparison of Average Newsprint Production Costs (1985) a/
(US$ per ton delivered to Western Europe)

| Cost Element | Western Canada | Eastern Canada | Sweden | Finland | Chile |
|---|---|---|---|---|---|
| Wood | 73 | 108 | 118 | 110 | 33 |
| Energy | 75 | 58 | 63 | 75 | 67 |
| Chemicals & other materials | 67 | 52 | 50 | 53 | 73 |
| Labor & Overheads | 127 | 125 | 60 | 76 | 86 |
| Capital charges | 52 | 38 | 37 | 39 | 60 |
| Transport | 50 | 45 | 25 | 30 | 45 |
| Total | 444 | 426 | 353 | 383 | 364 |

a/ 1985 data compared at 1985 exchange rates.

Source: Annex C, Table C.15.

14/ Until recently, pulp prices were fixed in dollars. However, first the strengthening, and subsequently the weakening of the dollar brought so much uncertainty to European producers and consumers that in most cases they have stopped using dollars, and in today's market, prices are usually fixed in the currency of the buyer.

The cost advantage is not as dramatic as with pulp, because wood is not such an important component in the manufacturing cost. However, the comparison shows that a country with low wood costs such as Chile, with no major distortion in the cost of any other factor of production, can deliver newsprint competitively to Europe. In fact, neither Chile nor any other developing country does ship newsprint to Europe. Chile's exports are primarily to Latin America where presumably the mill returns are higher. No other developing country exports any newsprint in any significant quantity.

## Price Trends

As with most other commodities, prices for forest products vary on a cyclical basis and, even at the same point in time, are often significantly different from one market to another. Major shifts in exchange rates add yet another variable to this equation. The largest trade in forest products is from Canada to the USA but because some of this trade is between branches of single companies, and because of the relatively low importance of freight, the prices found in this trade are not necessarily relevant to any other trading situation. Moreover, the volume of this trade is so great as to have a discernible effect on average world prices, even when this effect may not be evident in other world markets.

Notwithstanding these difficulties, it is of considerable importance to be able to trace the development of forest products prices over time. For this purpose, and in the absence of any relevant "international price" data, prices for major forest products imports into the UK have been chosen: first, because the UK is a major importer of all forest products; second, because it imports from most of the major producing regions so that international competition should be in evidence; and third, determining price trends in a single important currency may dampen some of the short-term exchange rate effects.

Price data for imports to the UK of softwood sawnwood, plywood, particleboard, bleached sulphate pulp and newsprint are contained in Annex C, Table C.16. The raw price information in UK pounds sterling per unit of product has been deflated using the IMF index of export unit values of manufactured goods, and then converted to an index with 1985 prices equal to 100. The main feature of this data is the very large cyclical fluctuations which are evident in the graphs shown in Figure 4.1. Between 1976/77 and 1980/81, for example, real prices of most forest products dropped by 40-45%. Also evident in the graphs (and statistically significant for sawnwood, pulp and newsprint) is a slight upward trend of prices, averaging between 1.8% and 2.4% annually over the past 20 years. This reflects real increases in costs as a result of more difficult access to suitable wood supplies, higher real energy costs and significant additional costs for pollution abatement. Only by building larger and more efficient plants has it been possible to maintain the growth in real prices at these low levels. These increases in both size and technological sophistication make it more difficult for new producers to enter the pulp and paper industry, particularly on the relatively small scale required by many developing countries.

Figure 4.1 - Real Price Trend Line

Mechanical Wood Products (1985 = 100)

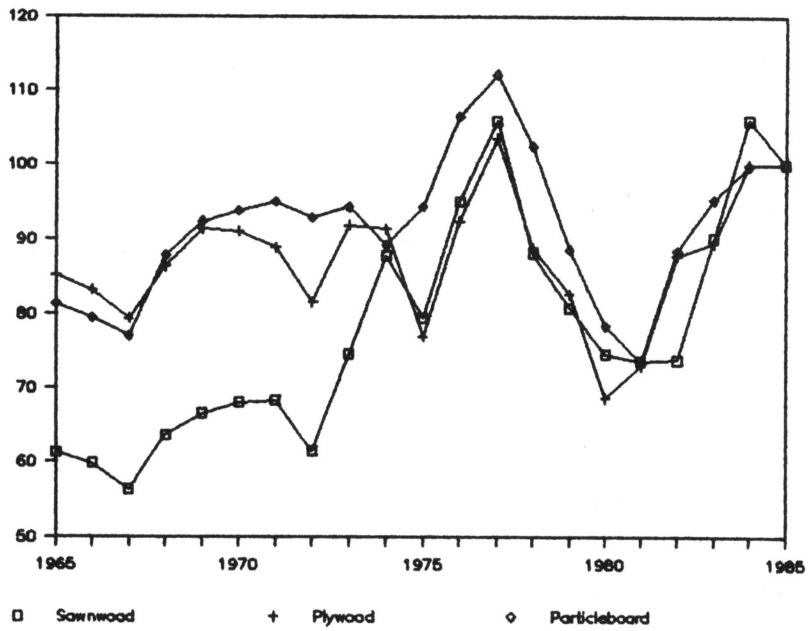


Bleached Pulp and Newsprint (1985 = 100)

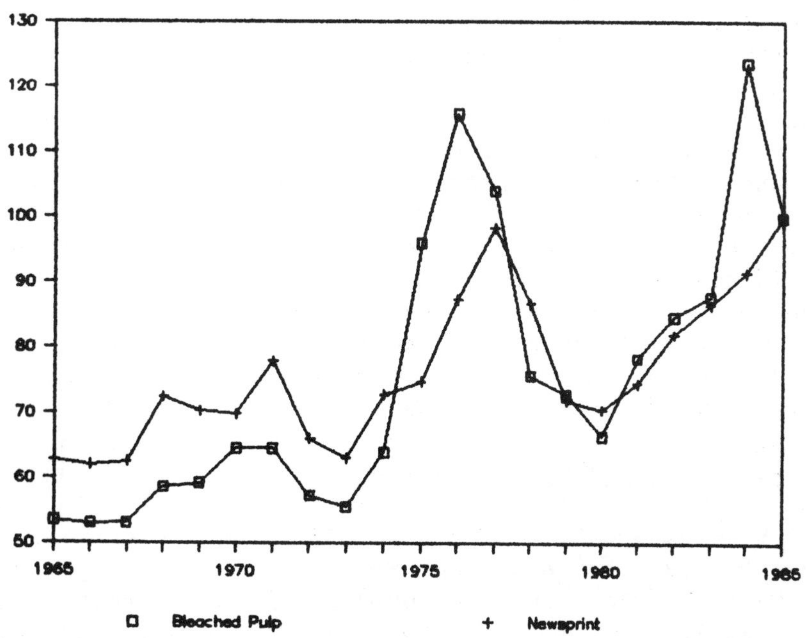

# V. POLICY ISSUES IN FOREST INDUSTRIES' DEVELOPMENT

Previous chapters have established that both domestic and export markets for forest products will grow, and that developing countries have the resources and the comparative advantage to compete successfully for an increasing share of these markets. Before attempting to quantify the extent of this increase, this chapter discusses some of the policy issues which must be addressed in order for any country to devise the appropriate framework to stimulate desired development in a particular situation.

The appropriate policy framework will differ substantially from country to country, and even amongst forest industries, from industry to industry. There are, nevertheless, some sector-specific issues, and considerable insight can be gleaned from an examination of actual developments in specific country situations. Of particular interest in this regard is the extent to which direct government intervention on the resource side may be appropriate. For the industries themselves, issues correspond more-or-less to those which may be encountered in the industrial sector as a whole, and the demonstrably correct role for government is to develop a policy and institutional package which will allow market signals to set clearly the pace of development activities at the industry level.

## Policy Issues and the Forest Resource

### Forest ownership

With important exceptions, most of the world's forests are owned by governments rather than by private individuals or corporations. For natural forests, this is entirely appropriate. Forests have many uses in addition to supplying wood to industrial users, such as the supply of fuelwood, recreation, hunting and trapping, watershed protection, nature conservation, etc., and this diversification is perhaps better controlled through public ownership. Moreover, most forests (including the northern forest of North America, the USSR and Scandinavia, and the vast areas of tropical rain forest) take 50-100 years to mature, which time-frame invites conflict between short-term profit maximization on the one hand, and long-term resource management for the public good on the other.

While more satisfactory than any alternative, public ownership of forests, and the general acceptance that because of the necessary long-term view, government involvement in forests is desirable, creates certain types of problems. In particular these relate to mechanisms for encouraging the establishment of new fast-growing forest plantations, and to systems for controlling the use and management of slower-growing natural forests.

### The establishment of new forests

Recent developments have led to the evolution of extremely fast growing clones of eucalyptus and other species which can reach pulpwood sizes in 8-10 years. Until recently, however, any investment in industrial plantations involved an investment for which no return could be anticipated for 15-25 years after the initial expenditure, and continuing costs (for maintenance and protection) throughout the intervening years. Moreover, the return on the initial investment, taking into account the time-value of money, was usually rather low. Under these circumstances, the establishment of industrial plantations was almost entirely the purview of governments, and particularly colonial governments which were impressed by the comparatively high rates of growth which could be found in tropical climates. In New Zealand, for example, make-work programs instituted during the depression led to the establishment of large areas of pine plantations, with no particular end-use in view. Similar plantations were established in many African colonies. Government railroads in some countries in Latin America and elsewhere planted significant areas of eucalyptus for fuelwood.

While much of this planting was to no identified end, it did provide valuable silvicultural and plantation management information. Samples of wood from early plantations were tested and found to be acceptable for the production of pulp and paper and some mechanical wood products. Particularly in such countries as Brazil, Portugal, Chile, New Zealand and South Africa, governments began to step up their rates of planting. In some countries, direct financial incentives were also introduced to encourage private planting.[15/] The results have been astounding. Brazil has established approximately 3 million hectares of industrial plantations, which when managed on a sustained yield basis will enable this country to produce as much industrial wood as all of the Scandinavian countries combined. With one of the most efficient industries in the world, Brazil ranks fifth in terms of pulp exports, all of which are produced from its relatively young industrial plantations. Chile has more than 1 million hectares of pine plantations, and New Zealand an approximately equal amount.

These accomplishments would not have been achieved without government intervention in the form of subsidies to private landowners for tree-planting. The subsidies themselves are not very much--in Chile, a total of US$50 million since 1974, in dollars of each year--for an industry which generated almost US$400 million in export revenues in 1984. Nor do they represent a major cost reduction in the end product--for pulp, for example, about US$6 per ton out of a total manufacturing cost of around US$200 per ton. Yet they stimulate an activity which would not otherwise take place, and which allows the subsequent development of a subsector with substantial economic input. It appears that proper economic accounting for

---

15/ The incentives differ somewhat. In Brazil, corporations and private land owners can invest in tree planting in lieu of paying taxes. In Chile, the Government reimburses up to 90% of the cost of tree-planting on private land. In both cases, subject to certain restrictions as to disposition of the mature trees, they are owned by the land-owner. Other countries, including the UK, for example, offer more limited tax incentives for tree planting.

industrial forestry has not yet been devised, and this is an issue which should be addressed by any country proposing to introduce a system of incentives to promote industrial afforestation.

Finally, with respect to relatively fast-growing industrial plantations, private ownership appears to be perfectly workable and indeed desirable. Private ownership may be on a large scale--such as the ownership of a block of plantation large enough to supply a modern industrial complex on a sustained-yield basis, or small scale, such as the group of small farmers in the Philippines who, with Bank support, established tree farms of fast-growing species to supply a pulp and paper mill on Mindanao.

Management of natural forests

As noted above, government ownership of natural forests which are being used to supply wood to private industry may be appropriate, but there are inevitably some problems at the interface between the forest-owner and the user of the resource. These usually fall into three distinct types: (i) the question of how long an operator may be conceded the right to operate in a certain area of forest (or in other situations, how much wood he may be allowed to remove); (ii) the question of establishing a mechanism for determining the value of wood as it is removed; and (iii) the question of replacing the resource. That there is no "best" answer to any of these questions is evidenced by the wide range of systems which are found to operate in both industrial and developing countries.

The size or time-frame of an industrial forest utilization concession is best determined in relation to the size and time frame of the associated industrial investment. If the concession is to extract timber in log form for immediate export, then 3-5 years might be appropriate as this is normally adequate time to recover the investment in roads and equipment. If, on the other hand, a major investment in pulp and paper production capacity is to be made, which would normally be amortized over 20-30 years, some assurance of a timber supply for at least that long a period, with some mechanism for further extension, is essential. Other types of operation (sawmilling, plywood production) would require security of wood supply for some intermediate period. The point is that whatever licensing or concession system is established to give rights to private industrialists to extract timber from state-owned forests must be flexible enough to adapt to the legitimate concerns of the industrialists regarding security of wood supply in relation to the type of investment.

The question of royalties or stumpage to be charged for timber removed from government-owned forests is the cause of even more difficulty. The notion of "replacement cost" is naturally attractive, but in slow-growing northern forests, and in tropical hardwood forests, compounding costs at any acceptable cost-of-capital rate for the 50-100 year period between planting trees and harvesting them gives figures which may be 5-10 times the stumpage figures actually being paid.[16] Were a particular government to attempt to levy such stumpages, the industry in that country would quickly be driven out of business. As with many other such situations, the result is a compromise. In Indonesia, for example, the government simply levies taxes and royalties which total about US$11 per cubic meter. This seems to be low enough (and is about the same in the Philippines, which competes in the same markets), because wood industries are more-or-less profitable in Indonesia. It could probably be raised to some extent, but a careful study of company profitability and comparative advantage would be a necessary precursor to such a move, in order to be able to counter the inevitable outcry which would follow any increase in royalties.

In British Columbia, Canada, a somewhat more sophisticated but essentially similar approach is adopted whereby stumpage fees for sawlogs are set on the basis of a formula developed by the Government, which determines the stumpage which would yield a certain rate of return on the invested capital of a typical (hopefully efficient) sawmiller. As the price of sawnwood varies, therefore, so does the stumpage charged. This system was accepted for years but has recently been challenged by sawmillers in the northwestern United States (who bid for cutting rights in open auction) as providing an unfair subsidy. As previously noted in Chapter IV, the US sawmillers have successfully argued this case to their Department of Commerce which has subsequently imposed a countervailing duty on the import of Canadian sawnwood.

The correct approach to determining the appropriate stumpage must clearly be a pragmatic but careful determination of how much revenue government believes it can extract for the use of the resource, what the market can bear, and what (if any) alternative uses for the resource may exist.

The final issue with respect to existing forest resources is the question of how to manage a forest area after it has been logged. The belief that governments will have a longer view than industrialists in this respect is one of the compelling justifications for government ownership and control of forest resources, but all too often, actual results fall short of desirable performance. In northern forests, logging is typically clear-felling. In Canada, 800,000 hectares are clear-felled each year. Of these areas, about 200,000 hectares are being planted or seeded artificially, another 300,000 are regenerating naturally, more-or-less satisfactorily, but the remaining 300,000 hectares is (in the words of the Canadian Forest Service) "being added to the regeneration backlog."

---

16/ Consider the simplified example of establishing a 1-hectare plantation of slow-growing natural forest at a cost of US$400, and waiting 50 years while it grew at a rate of 1.5 cubic meters per year. At the time of harvesting, at a 7% annual compound rate the US$400 would have grown to US$11,780, representing US$157 for each of the 75 cubic meters which could be harvested. Actual stumpages paid for such wood generally range from US$10-50 per cubic meter.

In tropical forests, the problems are somewhat different. Logging in these forests is typically on a selection system, whereby only "merchantable species" are removed, representing from perhaps 20% to 60% of the standing volume. Normally even merchantable species below a certain diameter (typically 30 cm) must be left, on the theory that these trees will grow larger, and can be cut at some later date. In Indonesia, for example, concession management is based on a 35-year rotation, which assumes that there will be commercial volumes of merchantable species 35 years after the original felling. In practice, there are many doubts about this system. First, without proper monitoring, operators may remove the smaller trees. Second, even if they are not removed illegally, they are frequently damaged during the initial logging operation, and do not recover. Finally, the character of the forest is so changed by the removal of much of the forest cover that there is no assurance that the younger trees will survive. This question is still subject to debate among foresters, but there are enough examples from inventories of logged-over areas some years after the initial logging to cast serious doubt on this approach to the management of tropical forests.

The point of this issue, in both northern and tropical forests, is the extent to which the first-time exploiter of the forests should be made responsible for re-establishing the resource in some form or other for a second exploitation, at some distant time in the future. There is clearly no natural economic reason for a concession-holder to undertake this work, and if it is required that he does, he must be tempted or coerced through a system of incentive or taxation. In any event this is an area where policy makers have an opportunity not only to ensure the effective operation of an industrial forestry operation in the short-term, but also to provide for continued operations into the long-term future.

## Coordination Between Forestry and Industry

Given the nature of industrial plantation projects, where the decision to plant trees may normally pre-date the start of mill construction by up to 15 years, it is not surprising that there are occasionally cases where plantations have been established, and then the mill which was to have used them does not get built. Malawi, for example, has established an excellent pine plantation for which there does not now seem to be any economic use. In other cases (Nigeria, Cameroon, Madagascar), proposed industries have not developed or sustained operations as planned, but in these cases there seem to be reasonable alternative uses for the wood which will be produced.

While all eventualities cannot be foretold, the issue here is that in planning industrial forestry operations, every effort should be made to analyse as wide a range of alternative end-uses as possible, in order to facilitate adjustment to a fallback position if this proves necessary. The viability and future economic competitiveness of the proposed industrial end-user should also be carefully evaluated prior to making large-scale investments in industrial plantations.

## Industry-Related Policy Issues

Apart from the specifically forestry-related issues as discussed above, many of the industrial-type issues are rather similar for forest industries as for other industries. They are nevertheless worth reiterating here, and the analysis of certain examples where the impact of specific policies (both right and wrong) can clearly be discerned in ensuing industrial development is illuminating.

### Trade policies

While the supposed justification of tariff protection and quantitative restrictions on imports may be to protect "infant industries," they can lead to the development of industries which are so inefficient that they could never compete in international markets. This can not only result in high prices for domestic consumers, but may divert efficient investments into areas which are financially rewarding for the investor, but economically inefficient.

A glaring example of the effects of a wrong trade policy can be found in the Indonesian paper industry. From the mid-1960s, the Government applied tariffs of up to 60% on paper imports, primarily to protect half a dozen small, inefficient and aging government-owned mills. No tariff was applied to pulp, and in the ensuing years, some thirty paper mills were built by private investors who could import pulp and other raw materials duty-free at world market prices, and sell paper internally at prices 60% higher. There was very little real value-added in these operations but it was financially so profitable that eventually too much capacity was built. By the early 1980s, although the tariff barriers were still in place, internal competition had brought paper prices in Indonesia down to levels approaching import parity. A number of the mills have closed down. Others have finally invested in badly-needed pulping capacity so that now, fifteen years later than necessary, the pulp and paper industry in Indonesia is developing an appropriately balanced structure. If this had taken place in a more timely manner, Indonesia could perhaps have by now developed a thriving export pulp and paper industry to supplement its plywood and sawnwood trade.

There are also examples of trade policies on the export side which have had a significant (positive and negative) effect on export competitiveness of developing countries. In real terms, trade in forest products has been growing at around 3.8% annually over the past decade, (somewhat faster than the total output of the industry), and trade from developing countries as a group has been growing at an even faster 4.8% annually. However, more than 90% of the growth in exports from developing countries can be attributed to just three countries: Brazil and Chile in pulp and paper, and Indonesia in the case of plywood. While one might look to the trade policies of Brazil and Chile to determine the reasons for their successes, a more important factor is the forestry policies.
In Indonesia, it was a trade policy which stimulated the rapid growth of the plywood exports: an effective ban on the export of logs. On the negative side, the log export business of Congo, for example, has dropped

sharply, at least in part because the Government established a timber marketing corporation through which all such logs had to be channelled. The result was log sales at less than world prices, and serious delays in effecting shipment. The Government is now apparently considering abolishing the marketing corporation. In other situations, particularly in countries with chronic foreign exchange shortages, companies may export successfully for a while, but then be barred from any access to the foreign exchange which they generate. If some of the foreign exchange is required to import spare parts or operating supplies, the company may have to cease operations, drying up all possibility of generating any more foreign exchange.

### Exchange rate policies

A realistic exchange-rate is crucial to the success of any export-oriented industry, and the experience of Chile shows the type of problems which can develop when exchange rates get seriously out of line. In the early 1980s, Chile's exchange rate was fixed vis-a-vis the dollar, and due to high rates of domestic inflation, the local currency soon became seriously overvalued in terms of dollars. Two of Chile's three large pulp and paper companies sold almost all of their output (newsprint and pulp) on world markets. While domestic production costs rose sharply with inflation, revenues remained flat. Reserves were drained, and both companies eventually went into bankruptcy. With Chile's subsequent economic reforms, the companies are recovering and now have new owners. Nevertheless, at a time when they could and should have been expanding to take advantage of growing markets and an expanding resource base, they were struggling to maintain their operations.

### Financial incentives and foreign investment

Governments may offer incentives such as tax holidays, accelerated depreciation and even direct grants to encourage investments. For smaller investments, such as for sawmills and even wood-based panel mills, these may have enough of an impact to encourage an investment which would not otherwise have been made. This may be appropriate as a policy tool to stimulate development, provided that a proper economic evaluation of the project is undertaken which includes all of the relevant costs, as well as the benefits (such as regional development) which may accrue outside the project itself. For larger investments, such as are required for a pulp and paper mill, these are unlikely to be significant enough to modify an investment decision.[17]

Attempts are frequently made to encourage investment in developing countries by companies active in forest industries in the traditional producing countries. For a variety of reasons, there has been

---

17/ Unless the grant is very large. In one case in Canada, for example, grants totalling over US$100 million led to the establishment of a plant which could never have been justified under normal economic criteria.

limited activity of this type. First, as previously discussed, most forest products are produced and consumed in the same country (or geographic region). Companies become familiar with local resource and market situations and do not usually have to look far afield for familiar project opportunities. Second, there is a perceived uncertainty about foreign situations. Forest industries are not usually particularly profitable and there is no incentive to take the unknown risks which may accompany an investment in an unknown environment. While for export-oriented projects there may be merit to the suggestion that the participation of a foreign company in the ownership is an indication of its worth, there is virtually no tradition of foreign investment in domestic-oriented forest-products companies, even in industrialized countries.

## Elements of a Recommended Policy Package

There is no policy package which can overcome serious technical, infrastructural or other drawbacks to the development of an efficient forest industry but in situations where these disadvantages do not exist, or are not too severe, the right policy package may help to stimulate development of efficient forest industries. Based on the foregoing review of good and bad policies now in effect, it is possible to provide at least the elements of a policy package which is likely to be successful.

First, any such package must be based on forestry legislation which clearly sets out the responsibilities for creation, management, exploitation and replacement of forests. Concession periods and wood pricing arrangements must be firm, and take into account the proposed end-use.

Second, trade policies must be such as to stimulate the development of efficient forest industries to supply domestic markets and, where appropriate, to export. No undue protection should be given to industries supplying the domestic market. If any such protection is to be given for "infant industries," it should not be particularly high and should be eliminated after a fixed period of time. Care must be taken to ensure that the protection does not allow industries to be established which, without it, would not be viable when settled down to steady operation. Access to foreign markets should be freely and directly available to individual companies. There may be advantages for companies from one country to join together and arrange their exporting as a group, but this should be voluntary to ensure that the system remains efficient. Further, for industries with a proven capacity to generate foreign exchange, safeguards must be introduced into any foreign exchange control scheme to ensure that the industries have access to enough of their foreign earnings to enable them to acquire necessary imported supplies, spare parts and other inputs required to remain competitive on world markets.

Third, exchange rates should be set (and where necessary adjusted) to ensure that importers and exporters are paying prices or receiving revenues which truly reflect the international value of the goods they are trading.

# VI. SUPPLY SCENARIOS FOR THE 1990s

Previous sections of this study have demonstrated that the demand for many forest products will increase between now and the end of the century; that there is adequate raw material to meet this demand; that for some products, developing countries can successfully compete in world markets; and that it is possible in these countries to establish the type of environment which will encourage efficient investments in forest industries.

This chapter attempts to predict, in general terms and on a broad regional and product basis, just how future demand might be supplied, and to what extent developing countries as a group, and on regional bases, can realistically hope to increase their share of supply.

## Resource-Constrained Supply

In terms of available forest resources, the major current deficit areas of the world are Western Europe and Japan. In 1985, these regions imported some US$30 billion worth of forest products, representing about one-half of world trade in these commodities. As noted in Chapter III, some increase in the supply of industrial wood is projected in these countries, but demand is also growing. With a regional breakdown of the demand forecast (Annex C, Table C.17) it is possible to prepare at least an indicative picture of the demand and (potential) supply for industrial roundwood in 2000. Table 6.1 presents such a summary.

According to this projection: (i) Western Europe and Japan will continue to be deficient in domestic wood supply to forest industries, at levels which are of the same order of magnitude as at present; and (ii) the rest of Asia will develop a serious deficit in industrial wood, and will thus have to become a major importer of logs or forest products. Overall, however, there will potentially be considerably more industrial wood than the projected requirement, as a result of large projected increases in supply from Latin America, North America and Eastern Europe/USSR. The actual supply situation will see demand and supply in balance, but it seems likely that rather than a proportional reduction in output from all areas, the deficit areas will produce to their maximum potential in order to minimize imports. Because of close links to Western Europe, Scandinavia will also cut wood at or near potential. Output from Africa, North and South America, Eastern Europe/USSR and Oceania will thus be reduced below the potential indicated in Table 6.1, in order to achieve the necessary balance.

There are a number of implications from this assessment. First, with a number of different countries and regions with potential supply surpluses, the potential exists for strong competition for markets for non-specialized products which can be produced from a variety of raw materials from different regions. Falling into this category are most grades of pulp and paper, reconstituted wood-based panels, and certain types of sawnwood. Not included in the category are products of tropical

Table 6.1

Current and Potential Regional Roundwood Surplus (Deficit)
(million cubic meters)

| | 1985 Surplus (Deficit) | Increased Roundwood Supply to 2000 | Roundwood Equivalent of Increased Demand to 2000 | 2000 Surplus (Deficit) |
|---|---|---|---|---|
| Africa | 3 | 23 | 15 | 11 |
| North America | 92 | 140 | 118 | 114 |
| Latin America | 1 | 112 | 39 | 74 |
| Western Europe | | | | |
| - Scandanavia | 64 | 15 | 7 | 72 |
| - Other | (94) | 45 | 59 | (108) |
| Asia | | | | |
| - Japan | (66) | 20 | 20 | (66) |
| - Other | (9) | 55 | 149 | (103) |
| Eastern Europe/USSR | 1 | 135 | 18 | 118 |
| Oceania | 8 | 30 | 3 | 35 |
| World | - | 575 | 428 | 147 |

Source: Derived from wood supply prognosis (Chapter III) and regional demand breakdown (Annex C, Table C.17).

hardwoods, and large-dimension and higher-quality sawnwood. Second, importing countries will continue to have the option, as they do now, to meet their deficits by importing any combination of logs, or semi-finished products, or finished products. These countries will effectively have the choice of installing additional capacity at home, and importing raw materials, or alternatively buying finished (or semi-finished) products on world markets. Third, countries/regions where the further development of forest industries is constrained by wood-supply will tend to direct resources to maximize value-added per scarce unit of resource.[18/]

## Tropical Hardwood Supply

As previously mentioned, the tropical hardwood market is one where competition is limited to those countries endowed with tropical forests and thus a supply scenario for this market must be developed

18/ Sweden, for example, where the annual production of industrial wood has been constrained over the past ten years, has significantly reduced exports of pulp and increased exports of paper to increase export earnings from the limited resource.

independently of other considerations. The following assumptions have been adapted from the Bank's commodity analysis for this purpose, although revisions have been introduced to develop projections of a suitable format for this report.

(a) World-wide demand for tropical hardwood and its products will increase by close to 1% annually between now and 2000. This figure is slightly higher than the projected demand for all hardwood sawnwood, implying a lower growth for the other hardwood grades.

(b) Asia will continue to be the main producing region, accounting in 2000 for 55% of total supply, compared with 26% for Latin America, and 16% for Africa.

(c) The degree of domestic processing will continue to increase, from current levels of around 70% to over 90% by 2000.

A supply/demand projection for tropical timbers, developed on this basis, is presented in Annex C, Table C.18, and summarized in Table 6.2.

Table 6.2

Tropical Hardwood Supply/Demand Projection (2000)
(million cubic meters)

| Region | Total Tropical Log Cut | Domestic Production | | Exports | | |
|---|---|---|---|---|---|---|
| | | Sawnwood | Plywood | Logs | Sawnwood | Plywood |
| Asia | 67.4 | 21.9 | 9.1 | 5.0 | 7.0 | 6.0 |
| Africa | 19.3 | 6.3 | 0.9 | 4.8 | 1.9 | 0.3 |
| Latin America | 31.8 | 13.8 | 1.7 | - | 2.5 | 0.6 |
| Oceania | 2.1 | 0.3 | 0.1 | 1.5 | 0.1 | - |
| Total | 120.6 | 42.3 | 12.0 | 11.3 | 11.5 | 6.9 |

Source: Annex C, Table C.18.

## Cost-efficient Supply

Apart from those areas which are projected to be deficit in wood, and the special case of tropical hardwoods, it may be supposed that in the absence of any other distortions, the balance of the world's demand for forest products in the future will be met by the most cost-efficient producers. This supposition is examined below, on a product by product basis.

Sawnwood

Only about 10% of all sawnwood produced is traded internationally. Sawnwood for ordinary construction uses can be made relatively efficiently on a relatively small scale and where raw material is available, the most cost efficient supply in most cases is domestic supply. Thus, as might be expected, the regions which are not deficit in wood, show high levels of domestic self-sufficiency in sawnwood, as shown in Table 6.3.

Table 6.3

Self-Sufficiency in Sawnwood Supply for Selected Regions (1985)
(percent)

| | Hardwood | Softwood |
|---|---|---|
| Africa | 102 | 35 |
| Asia | 108 | 89 |
| North America | 98 | 106 |
| Latin America | 103 | 95 |
| USSR/Eastern Europe | 100 | 106 |
| Oceania | 89 | 83 |

Out of the approximately 65 forested countries within these regions, all but 10 produce substantially all of their own sawnwood. With further increases in forest resources projected, it is reasonable to assume that these regions will continue to maintain 100% self-sufficiency in terms of domestic supply of sawnwood into the foreseeable future.[19/]

In international trade, consumption in the two main importing regions--Japan and Western Europe--is projected to remain more or less constant and with slight increases in projected domestic roundwood availability, imports may be expected to decline. There may be opportunities for competitive producers in Latin America to displace traditional suppliers (primarily North America and Scandinavia for coniferous sawnwood, and forests with tropical forests for hardwood), but because of real or perceived quality limitations, the degree of penetration is not likely to be great. The main new growth area for sawnwood are China and the other wood deficit countries of Asia, where imports can be expected to increase by upwards of 20 million cubic meters annually through the end of the decade. With no prejudices or traditions, these developing markets

19/ The USA is actually only 77% self-sufficient in sawnwood, and the balance (almost all of which is imported from Canada) represents more imports than any other country. US imports from Canada are mainly construction grade timber and competitive producers who can produce this grade could displace Canadian suppliers to some extent.

are likely the best opportunities for additional sawnwood capacity in the traditional producing areas of North America and the USSR, and for new producers in Latin America and Oceania.

Panels

Veneer and plywood are traded internationally more than sawnwood (about 18% of all production), partly because log requirements are more demanding, and fewer countries have logs, and partly because even the simplest plywood plant requires more capital and technology than a simple sawmill. Small countries with small markets may not be able to justify the investment. Also, in most applications sawnwood is a feasible alternative to panels: according to FAO statistics, 27 countries are presently consuming no wood-based panels at all. Particleboard and fiberboard are traded rather less, and usually only over relatively short distances, because their relatively low value-to-weight ratio makes competition susceptible to freight costs. Thus, except for the special case of tropical hardwood veneers and plywood, even the wood-poor areas of Europe and Japan manage to be virtually self-sufficient in their demands for panels.

For these reasons, a marketing strategy for developing countries for reconstituted wood-based panels should concentrate on regional markets, and here there are some interesting opportunities. In Asia, for example, particleboard consumption is projected to double by the end of the century, to over 4.5 million cubic meters annually. The keys to success in this market are low-cost wood, resin availability at close to international prices, market opportunities to allow economies of scale in plant construction, and access to these markets at reasonable freight rates.

Pulp and paper

Although paper is produced in over 90 countries, levels of self-sufficiency are generally lower for this commodity than for any other forest product, and about 40 million tons--21% of all paper production--is traded internationally. About 50% of this trade is within Europe, and a further 25% between Canada and the USA. In 1985, only about 4% of all pulp and paper exports originated in developing countries, but this share has doubled in the past ten years.

The demand projections developed in Chapter II suggested that annual world demand for paper would increase by about 104 million tons between 1985 and 2000, which corresponds to a requirement for new capacity of about 114 million tons based on historical rates of capacity utilization. The objective here is to develop a plausible scenario (or scenarios) for determining the most likely geographical location for this paper production capacity, and the pulp capacity required to support it. For this analysis, the following approach has been developed:

(a) Recent trends in self-sufficiency in paper production have been determined, and projected to continue to the end of the century. This implies that industrial economies, by regions, will be at least self-sufficient in paper production, and will import pulp as required to make this paper;

(b) In the wood-deficit areas, regional pulp production will increase to the level constrained by projected wood supplies, after allowing for wood use for other industrial purposes; and

(c) The remaining pulp deficit will be allocated to the wood-surplus regions taking into consideration: (i) traditional supply patterns; (ii) historical market share and changing trends in this share; (iii) geographical location vis-a-vis the deficit areas; and (iv) comparative cost advantage.

Supply scenario

From the above analysis of the likely supply scenario of the three main product groupings, a balanced supply/demand model for the world's forest products has been prepared. The detailed model is presented in tabular form in Annex C, Table C.19, and summarized in Table 6.4.

Table 6.4

Forest Products Supply Model
(%)

| | Self Sufficiency in the Supply of: | | | | | |
|---|---|---|---|---|---|---|
| | Sawnwood | | Panels | | Paper | |
| | 1985 | 2000 | 1985 | 2000 | 1985 | 2000 |
| Industrial Countries | 100 | 103 | 95 | 97 | 103 | 105 |
| Developing Countries: | | | | | | |
| Africa | 66 | 70 | 91 | 105 | 38 | 40 |
| Asia & Oceania | 103 | 90 | 140 | 114 | 83 | 80 |
| Latin America | 99 | 110 | 108 | 100 | 91 | 95 |
| Group Total | 99 | 93 | 126 | 110 | 83 | 83 |

In most respects, this scenario projects that future supply patterns will be similar to current experience. The major shift projected is a significant drop in the self-sufficiency level in Asia for both sawnwood and wood-based panels. This is not to say production will decline in this region; in fact it will increase by about 3% annually for sawnwood and by more than 5% for panels. However, the scenario assumes that with flat or falling demand in the industrial countries which will enhance supply possibilities from traditional producing areas, and the possibilities for very cheap sawnwood from Latin America, capacity additions in Asia will not keep pace with growing demand.

## Investment Implications

At the conclusion of Chapter II, it was estimated that the world's forest industry will have to invest about US$14 billion annually for capacity additions to supply growing world demand for forest products. According to the supply scenarios discussed above, the product and regional breakdown of these investments would be roughly as shown in Table 6.5, based on calculations detailed in Annex C, Table C.20.

Table 6.5

Average Annual Investment Requirements for Forest Industries
(US$ million/year for 1985-2000)

| | Sawnwood | Panels | Pulp & Paper | Total |
|---|---|---|---|---|
| Industrial Countries | 160 | 380 | 8,430 | 8,970 |
| Developing Countries: | | | | |
| Africa | 110 | 20 | 50 | 180 |
| Asia & Oceania | 470 | 150 | 2,540 | 3,160 |
| Latin America | 260 | 50 | 1,580 | 1,890 |
| Subtotal | 840 | 220 | 4,170 | 5,230 |
| World | 1,000 | 600 | 12,600 | 14,200 |

Source: Annex C, Table C.20.

The global sum of all gross investments in all sectors in 1985 was some US$2,500 billion and projected investments in forest industries represent about 0.6% of this figure. This is not unreasonable on the basis that forest industries contribute about 0.8% to world GNP and thus merit at least a comparable share of investments. However, based on the supply scenario outlined above, there are significant differences in the share of forest investments to total investments on a regional basis. While investments in the forest industries of industrial countries would amount to a fairly representative 0.4% of total investments, the proportion would range from 1.6% in Latin American countries, to 1.0% in Asia and 0.3% in Africa. The high level of investment indicated in Latin America may be constrained by capital availability, which would modify the scenario outlined above.

## ANNEX A

### Glossary of Products and Processes 1/

ALKALINE PULPING PROCESSES Usually refers to either the KRAFT or SODA pulping process in which the main pulping chemical is caustic soda.

ACID SULPHITE PULPING PROCESS A process for producing SULPHITE pulp, mainly used for preparing DISSOLVING PULP.

BAGASSE PULP Pulp made from BAGASSE which is sugar-cane residue after sugar has been extracted. Pulp is usually prepared by the SODA process but a number of MECHANICAL and SEMI-CHEMICAL processes have been developed in order to allow the production of higher-yield pulps. Bagasse pulp is SHORT-FIBERED.

BASIS WEIGHT The weight per unit area of paper, usually measured in grams per square meter (g/m2).

BROAD-LEAVED See NON-CONIFEROUS.

CHEMI-THERMOMECHANICAL PULP Pulp produced by a MECHANICAL process in which CONIFEROUS or NON-CONIFEROUS wood particles are softened by pre-heating under pressure, and with chemicals, prior to a pressurized refining stage. It may be bleached or unbleached. The chemical pretreatment provides a pulp with better strength properties than THERMOMECHANICAL PULP.

COATED PAPER Printing and writing paper which has been coated on one or both sides with coating materials such as clay (kaolin), barium sulphate, gypsum or zinc oxide.

COMPRESSED FIBERBOARD A type of FIBERBOARD with a relatively high degree of pressing and a relatively high density. Typically HARDBOARD.

CONIFEROUS Refers to wood derived from trees classified botanically as "Gymnospermae", e.g. Abies, Agathis, Araucaria, Cedrus, Larix, Picea, Pinus, etc. The term SOFTWOOD usually refers to timber in this group. Coniferous pulps are usually LONG-FIBERED.

CORRUGATING MEDIUM See FLUTING.

---

1/ Many of the definitions contained herein are from "Classification and Definitions of Forest Products". FAO Forestry Paper No. 32. Food and Agriculture Organization of the United Nations. Rome, 1982.

DISSOLVING PULP Highly bleached chemical pulp (SULPHATE, SODA or SULPHITE) from CONIFEROUS or NON-CONIFEROUS wood, rags, cotton linters, etc., of special quality, with a very high alpha-cellulose content (usually 90% or more). These pulps are used principally as a source of cellulose in the manufacture of products such as man-made fibers, cellulosic plastic materials, lacquers, explosives, etc.

FIBERBOARD Sheet material usually exceeding 1.5 mm in thickness, manufactured from fibers of wood (or other ligno-cellulosic materials) with the primary bond deriving from the felting of the fibers and their inherent adhesive properties. Bonding materials and other chemicals may be added. Fiberboard is usually flat-pressed, but may be molded. The density of the board varies with the degree of compresssion. A relatively light pressing is used for such NON-COMPRESSED FIBREBOARDs as INSULATING BOARD. High pressures are used for COMPRESSED FIBERBOARD or HARDBOARD. See also MEDIUM DENSITY FIBERBOARD.

FOLDING BOXBOARD Paperboard with good stiffness, scoring and folding characteristics. These paperboards are made from pulp, waste paper, or any combination of these.

FLUTING Paper or paperboard used mainly as a corrugating medium in the production of corrugated board. It is typically produced wholly or principally from semi-chemical wood pulp.

HARDBOARD A type of COMPRESSED FIBERBOARD with a density usually over 0.8 g/cm3.

HARDWOOD See NON-CONIFEROUS.

INSULATING BOARD Fiberboard with a density usually not more than 0.35 g/cm3, sometimes known as SOFTBOARD.

KRAFT PULP Frequently used to describe pulp made by the SULPHATE process

LINERBOARD Paperboard made either from sulphate pulp (bleached or unbleached) or principally from waste paper, used as facing material on corrugated or solid paper or paperboard boxes and containers.

LONG-FIBERED PULP A pulp where the average fiber length is relatively long. Characteristic properties of paper produced from long-fibered pulp are relatively high bursting and tearing strength.

MECHANICAL PULP Pulp prepared either by grinding wood against a stone or by refining chips. The yield is very high but the strength is rather low. The pulp may be unbleached or bleached.

MEDIUM DENSITY FIBERBOARD Dry process FIBERBOARD with a density of about 0.7 g/cm3. The chip raw material is ground into fiber under steam pressure before the addition of resin. The resulting homgeneity of texture gives medium density fiberboard (MDF) desirable qualities as regards machining.

NEWSPRINT Paper of the type used mainly for the printing of newspapers. It usually contains at least 65% MECHANICAL or THERMOMECHANICAL pulp, and has a BASIS WEIGHT not less than 40 or more than 57 g/m2.

NON-COMPRESSED FIBERBOARD A type of FIBERBOARD with a relatively low degree of pressing and a relatively low density. Typically INSULATING BOARD.

NON-CONIFEROUS Refers to wood derived from trees classified botaincally as "Angiospermae", e.g. Betula, Casuarina, Dipterocarpus, Eucalyptus, Fagus, Khaya, Populus, Quercus, Shorea, Swietenia, Tectona, Terminalia, etc. The terms HARDWOOD and BROAD-LEAVED usually refer to timbers in this group. See also CONIFEROUS. Non-coniferous wood are usually SHORT FIBERED.

ORIENTED STRUCTURAL BOARD A special type of platen pressed particle board in which the particles are all oriented in the same direction, thereby improving the strength of the board in the direction in which the particles are oriented.

PARTICLEBOARD Flat-pressed or molded panels manufactured from particles of wood or other ligno-cellulosic materials bonded by a suitable (usually organic) binder in the presence of heat and pressure. Particleboard may also be made by extruding the materials into a mold under heat and pressure. The density of particleboard is usually within the range 0.6 to 0.8 g/cm2. See also ORIENTED STRUCTURAL BOARD and WAFERBOARD.

PEELED VENEER VENEER produced by applying a knife tangentially to a rotating log. See also SLICED veneer.

PLYWOOD A panel consisting of an assembly of veneer sheets bonded together with the direction of the grain in alternative plies generally at right angles. Plywood is usually flat pressed but may be molded.

PRINTING AND WRITING PAPER Paper other than newsprint suitable for a wide variety of printing and other graphic purposes, made from a variety of furnishes and with various finishes. It may be COATED or uncoated.

PULPING PROCESSES Any of a variety of processes used to convert cellulosic materials (usually wood) to a pulp suitable for making paper. See also ALKALINE PULPING PROCESSES, ACID SULPHITE PULPING PROCESS, MECHANICAL PULP, THERMOMECHANICAL PULP, SEMI-CHEMICAL PULP, and CHEMI-THERMOMECHANICAL PULP.

SAWNWOOD Includes wood simply sawn lengthwise, or produced by a profile chipping process. With few exceptions, sawnwood exceeds 5 mm in thickness.

SEMI-CHEMICAL PULP A pulp produced by treating wood with a moderate amount of chemical (usually neutral SULPHITE) before grinding or refining. The resulting pulp is normally between MECHANICAL and CHEMICAL PULP in terms of both strength and yield.

SLICED VENEER VENEER produced by slicing squared billets of wood along the grain. This method of producing veneer is usually reserved for high value decorative species. See also PEELED veneer.

SODA PULP Pulp made by cooking chips in a pressure vessel in the presence of a cooking liquor of sodium hydroxide solution.

SOFTBOARD See INSULATING BOARD.

SOFTWOOD See CONIFEROUS.

STRAW PULP Pulp made from straw, usually either wheat straw or rice straw pulped with the SODA process. Straw pulps are SHORT-FIBERED.

SULPHATE PULP Pulp made by cooking chips in a pressure vessel in the presence of a cooking liquor of sodium hydroxide and sodium sulphide solution.

SULPHITE PULP Pulp made by cooking chips in a pressure vessel in the presence of a bisulphite cooking liquor. Bisulphites such as ammonium, calcium, magnesium and sodium are frequently used.

THERMO-MECHANICAL PULP Pulp produced by a MECHANICAL process in which CONIFEROUS or NON-CONIFEROUS wood particles are softened by pre-heating under pressure prior to a pressurized refining stage. It may be bleached or unbleached.

VENEER Thin layer or sheet of wood of uniform thickness (usually 5 mm or less), SLICED or PEELED, for use in making PLYWOOD, or for veneering furniture, etc.

WAFERBOARD A special type of platen pressed particleboard made up of long particles "wafers", thereby improving the strength of the board.

WOOD-CONTAINING PAPER Printing and writing papers containing 10% or more MECHANICAL or THERMOMECHANICAL PULP.

WOOD-FREE PAPER Printing and writing papers containing less than 10% MECHANICAL or THERMOMECHANICAL pulp.

# ANNEX B

## List of Countries Included in Economic Classes and Regions [1/]

### Developed Market Economies

#### North America
Canada, United States

#### Western Europe
Austria, Belgium-Luxembourg, Denmark, Finland, France, Germany (Federal Republic), Greece, Iceland, Ireland, Italy, Malta, Netherlands, Norway, Portugal, Spain, Sweden, Switzerland, United Kingdom, Yugoslavia

#### Oceania
Australia, New Zealand

#### Other Developed Market Economies
Israel, Japan, South Africa

### Developing Market Economies

#### Africa
Algeria, Angola, Benin, Botswana, Burkina Faso, Burundi, Cameroon, Cape Verde, Central African Republic, Chad, Congo, Djibouti, Equatorial Guinea, Egypt, Ethiopia, Gabon, Gambia, Ghana, Guinea Bissau, Guinea, Ivory Coast, Kenya, Lesotho, Liberia, Libya, Madagascar, Malawi, Mali, Mauritania, Mauritius, Morocco, Mozambique, Niger, Nigeria, Reunion, Rwanda, Sao Tome and Principe, Senegal, Sierra Leone, Somalia, Sudan, Swaziland, Tanzania, Togo, Tunisia, Uganda, Zaire, Zambia, Zimbabwe

#### Latin America
Argentina, Bahamas, Barbados, Belize, Bolivia, Brazil, Chile, Colombia, Costa Rica, Cuba, Dominica, Dominican Republic, Ecuador, El Salvador, French Guiana, Guadeloupe, Guyana, Haiti, Honduras, Jamaica, Martinique, Mexico, Netherlands Antilles, Nicaragua, Panama, Paraguay, Peru, Suriname, Trinidad and Tobago, Uruguay, Venezuela

For statistical convenience, where the designation "industrial countries" is used, it includes all of the Developed Market Economies in this listing, together with the Centrally Planned Economies of Europe, and the USSR.

---

1/ From: "Yearbook of Forest Products". FAO, Rome, 1986.

Pakistan, Philippines, Qatar, Kingdom of Saudi Arabia, Singapore, Sri Lanka, Thailand, Turkey, Yemen Democratic Republic

Other Developing Market Economies

Fiji, French Polynesia, New Caledonia, Papua New Guinea, Samoa, Solomon Islands, Tonga, Vanuatu

Centrally Planned Economies

Asia

China, Democratic People's Republic of Korea, Kampuchea Democratic, Mongolia, Viet Nam

Eastern Europe and USSR

Albania, Bulgaria, Czechoslovakia, German Democratic Republic, Hungary, Poland, Romania, USSR

# ANNEX C

## Supplementary Tables

### Table C.1 - Development of Sawnwood Demand, 1955-1985

| | Amount (millions of cu m 3-year averages) | | | | Average Annual Increase | | | Period Demand/Income Elasticity | | |
|---|---|---|---|---|---|---|---|---|---|---|
| | 1955 | 1965 | 1975 | 1985 | 1955-65 | 1965-75 | 1975-85 | 1955-65 | 1965-75 | 1975-85 |
| **DEVELOPING COUNTRIES** | | | | | | | | | | |
| Production | 31 | 41 | 65 | 103 | 2.8% | 4.9% | 4.6% | | | |
| - hardwood | 19 | 24 | 41 | 64 | 2.7% | 5.2% | 4.6% | | | |
| - softwood | 12 | 16 | 25 | 39 | 2.9% | 4.4% | 4.7% | | | |
| Imports | 4 | 4 | 7 | 11 | -0.9% | 5.8% | 5.2% | | | |
| - hardwood | 2 | 1 | 3 | 4 | -2.8% | 8.6% | 3.8% | | | |
| - softwood | 3 | 3 | 4 | 8 | -0.1% | 4.5% | 6.0% | | | |
| Exports | 3 | 4 | 8 | 10 | 3.4% | 7.0% | 2.3% | | | |
| - hardwood | 1 | 3 | 6 | 8 | 5.3% | 9.4% | 2.8% | | | |
| - softwood | 1 | 1 | 2 | 2 | 0.7% | 0.9% | 0.3% | | | |
| Apparent Consumption | 32 | 41 | 65 | 105 | 2.3% | 4.8% | 4.9% | 0.48 | 0.85 | 1.14 |
| - hardwood | 19 | 23 | 37 | 59 | 2.1% | 4.8% | 4.9% | 0.44 | 0.86 | 1.12 |
| - softwood | 14 | 18 | 28 | 45 | 2.6% | 4.7% | 5.1% | 0.53 | 0.83 | 1.16 |
| **INDUSTRIAL COUNTRIES** | | | | | | | | | | |
| Production | 278 | 333 | 356 | 361 | 1.8% | 0.7% | 0.1% | | | |
| - hardwood | 55 | 66 | 63 | 52 | 1.7% | -0.4% | -1.8% | | | |
| - softwood | 223 | 267 | 294 | 308 | 1.8% | 1.0% | 0.5% | | | |
| Imports | 29 | 43 | 53 | 73 | 4.2% | 2.1% | 3.3% | | | |
| - hardwood | 3 | 4 | 8 | 9 | 3.8% | 5.6% | 1.0% | | | |
| - softwood | 26 | 39 | 45 | 65 | 4.2% | 1.6% | 3.6% | | | |
| Exports | 31 | 44 | 53 | 76 | 3.4% | 1.9% | 3.6% | | | |
| - hardwood | 3 | 4 | 4 | 4 | 1.4% | 1.9% | -0.4% | | | |
| - softwood | 28 | 41 | 49 | 72 | 3.6% | 1.9% | 3.9% | | | |
| Apparent Consumption | 275 | 332 | 356 | 358 | 1.9% | 0.7% | 0.1% | 0.40 | 0.18 | 0.02 |
| - hardwood | 55 | 67 | 66 | 57 | 1.9% | .0% | -1.6% | 0.39 | -0.01 | -0.61 |
| - softwood | 220 | 265 | 290 | 302 | 1.9% | 0.9% | 0.4% | 0.40 | 0.23 | 0.15 |
| **WORLD TOTALS** | | | | | | | | | | |
| Production | 309 | 373 | 422 | 464 | 1.9% | 1.2% | 1.0% | | | |
| - hardwood | 74 | 90 | 103 | 116 | 2.0% | 1.4% | 1.2% | | | |
| - softwood | 235 | 283 | 318 | 348 | 1.9% | 1.2% | 0.9% | | | |
| Imports | 33 | 47 | 60 | 85 | 3.6% | 2.4% | 3.5% | | | |
| - hardwood | 5 | 6 | 10 | 12 | 2.0% | 6.3% | 1.8% | | | |
| - softwood | 28 | 42 | 50 | 72 | 3.9% | 1.8% | 3.9% | | | |
| Exports | 34 | 48 | 61 | 86 | 3.4% | 2.4% | 3.5% | | | |
| - hardwood | 5 | 6 | 11 | 12 | 2.8% | 5.6% | 1.6% | | | |
| - softwood | 30 | 42 | 50 | 73 | 3.5% | 1.9% | 3.8% | | | |
| Apparent Consumption | 307 | 372 | 421 | 463 | 1.9% | 1.2% | 1.0% | 0.40 | 0.30 | 0.33 |
| - hardwood | 74 | 90 | 103 | 116 | 1.9% | 1.4% | 1.2% | 0.40 | 0.34 | 0.41 |
| - softwood | 233 | 283 | 318 | 347 | 1.9% | 1.2% | 0.9% | 0.40 | 0.28 | 0.31 |

GDP growth for elasticity calculations

| | 1955-65 | 1965-75 | 1975-85 |
|---|---|---|---|
| Developing countries | 4.9% | 5.6% | 4.4% |
| Industrial countries | 4.8% | 3.9% | 2.6% |
| World | 4.8% | 4.2% | 2.9% |

Table C.2 - Development of Wood-based Panel Demand, 1955-1985

| | Amount (millions of cu m 3-year averages) | | | | Average Annual Increase | | | Period Demand/Income Elasticity | | |
|---|---|---|---|---|---|---|---|---|---|---|
| | 1955 | 1965 | 1975 | 1985 | 1955-65 | 1965-75 | 1975-85 | 1955-65 | 1965-75 | 1975-85 |
| DEVELOPING COUNTRIES | | | | | | | | | | |
| Production | 0.8 | 3.2 | 9.3 | 19.6 | 15.5% | 11.1% | 7.8% | | | |
| - veneer | 0.2 | 0.6 | 1.1 | 1.7 | 10.3% | 5.8% | 4.7% | | | |
| - plywood | 0.5 | 2.0 | 5.7 | 12.4 | 15.1% | 11.1% | 8.2% | | | |
| - particleboard | .0 | 0.2 | 1.5 | 3.2 | 40.1% | 19.4% | 8.2% | | | |
| - comp fibreboard | .0 | 0.3 | 0.9 | 1.9 | 21.5% | 12.5% | 8.5% | | | |
| - non-comp f'board | .0 | 0.1 | 0.2 | 0.3 | 51.3% | 3.8% | 3.4% | | | |
| Imports | 0.2 | 0.3 | 1.3 | 3.4 | 7.4% | 14.0% | 10.3% | | | |
| - veneer | .0 | .0 | 0.2 | 0.3 | 3.3% | 30.3% | 5.6% | | | |
| - plywood | 0.1 | 0.2 | 0.8 | 2.6 | 6.9% | 13.4% | 12.8% | | | |
| - particleboard | .0 | .0 | 0.1 | 0.3 | 6.2% | 15.7% | 8.9% | | | |
| - comp fibreboard | .0 | 0.1 | 0.1 | 0.1 | 11.8% | 8.9% | -0.8% | | | |
| - non-comp f'board | .0 | .0 | 0.1 | 0.1 | 8.9% | 9.8% | 0.2% | | | |
| Exports | 0.1 | 1.2 | 4.0 | 7.4 | 23.1% | 13.2% | 6.2% | | | |
| - veneer | .0 | 0.3 | 0.6 | 1.0 | 20.3% | 7.8% | 5.2% | | | |
| - plywood | 0.1 | 0.8 | 3.2 | 6.1 | 24.3% | 14.2% | 6.7% | | | |
| - particleboard | .0 | .0 | 0.1 | 0.1 | 21.1% | 17.0% | -4.4% | | | |
| - comp fibreboard | .0 | .0 | 0.1 | 0.3 | 26.9% | 22.2% | 6.2% | | | |
| - non-comp f'board | .0 | .0 | .0 | .0 | 26.5% | 17.6% | 1.7% | | | |
| Apparent Consumption | 0.8 | 2.4 | 6.5 | 15.6 | 11.9% | 10.4% | 9.1% | 2.45 | 1.86 | 2.10 |
| - veneer | 0.2 | 0.4 | 0.7 | 1.1 | 6.2% | 6.6% | 4.6% | 1.27 | 1.18 | 1.05 |
| - plywood | 0.5 | 1.4 | 3.3 | 8.9 | 10.5% | 9.2% | 10.6% | 2.16 | 1.65 | 2.43 |
| - particleboard | .0 | 0.3 | 1.5 | 3.5 | 27.0% | 19.2% | 8.9% | 5.57 | 3.42 | 2.06 |
| - comp fibreboard | 0.1 | 0.3 | 0.8 | 1.8 | 18.8% | 10.9% | 7.9% | 3.88 | 1.95 | 1.81 |
| - non-comp f'board | .0 | 0.1 | 0.2 | 0.3 | 28.1% | 4.9% | 2.7% | 5.80 | 0.87 | 0.62 |
| INDUSTRIAL COUNTRIES | | | | | | | | | | |
| Production | 19.7 | 43.8 | 79.9 | 91.7 | 8.3% | 6.2% | 1.4% | | | |
| - veneer | 1.8 | 2.2 | 2.7 | 3.0 | 1.7% | 2.2% | 1.1% | | | |
| - plywood | 9.9 | 21.5 | 30.7 | 32.5 | 8.1% | 3.6% | 0.6% | | | |
| - particleboard | 0.5 | 8.6 | 30.9 | 42.1 | 34.0% | 13.6% | 3.1% | | | |
| - comp fibreboard | 1.7 | 4.0 | 7.1 | 7.4 | 9.0% | 5.9% | 0.4% | | | |
| - non-comp f'board | 5.9 | 7.4 | 8.5 | 6.7 | 2.4% | 1.3% | -2.4% | | | |
| Imports | 1.8 | 5.3 | 12.3 | 15.7 | 11.4% | 8.8% | 2.5% | | | |
| - veneer | 0.3 | 0.7 | 1.2 | 1.6 | 9.2% | 6.3% | 2.6% | | | |
| - plywood | 0.8 | 2.3 | 5.2 | 5.4 | 11.6% | 8.4% | 0.4% | | | |
| - particleboard | 0.1 | 0.8 | 4.1 | 6.6 | 30.6% | 17.1% | 4.9% | | | |
| - comp fibreboard | 0.4 | 0.9 | 1.1 | 1.2 | 7.8% | 2.4% | 0.9% | | | |
| - non-comp f'board | 0.3 | 0.6 | 0.7 | 0.9 | 8.2% | 1.6% | 3.3% | | | |
| Exports | 1.8 | 4.5 | 9.2 | 11.8 | 9.5% | 7.4% | 2.5% | | | |
| - veneer | 0.2 | 0.3 | 0.6 | 0.9 | 6.4% | 6.7% | 4.7% | | | |
| - plywood | 0.8 | 1.7 | 2.4 | 2.6 | 7.5% | 3.7% | 0.7% | | | |
| - particleboard | 0.1 | 0.9 | 4.2 | 6.2 | 28.1% | 17.2% | 3.8% | | | |
| - comp fibreboard | 0.5 | 1.0 | 1.2 | 1.2 | 7.5% | 2.2% | -0.1% | | | |
| - non-comp f'board | 0.3 | 0.7 | 0.8 | 0.9 | 8.8% | 1.3% | 1.3% | | | |
| Apparent Consumption | 19.7 | 44.6 | 83.0 | 95.6 | 8.5% | 6.4% | 1.4% | 1.79 | 1.64 | 0.56 |
| - veneer | 1.9 | 2.5 | 3.3 | 3.7 | 2.6% | 2.8% | 1.0% | 0.55 | 0.73 | 0.39 |
| - plywood | 9.8 | 22.2 | 33.5 | 35.3 | 8.5% | 4.2% | 0.5% | 1.78 | 1.08 | 0.21 |
| - particleboard | 0.4 | 8.6 | 30.8 | 42.5 | 34.4% | 13.6% | 3.3% | 7.24 | 3.48 | 1.28 |
| - comp fibreboard | 1.6 | 3.9 | 7.0 | 7.4 | 9.1% | 6.0% | 0.6% | 1.91 | 1.54 | 0.23 |
| - non-comp f'board | 5.8 | 7.4 | 8.4 | 6.7 | 2.3% | 1.3% | -2.2% | 0.49 | 0.34 | -0.85 |

Table C.2 (continued)

| | Amount (millions of cu m 3-year averages) | | | | Average Annual Increase | | | Period Demand/Income Elasticity | | |
|---|---|---|---|---|---|---|---|---|---|---|
| | 1955 | 1965 | 1975 | 1985 | 1955-65 | 1965-75 | 1975-85 | 1955-65 | 1965-75 | 1975-85 |
| WORLD TOTALS | | | | | | | | | | |
| Production | 20.5 | 47.0 | 89.2 | 111.2 | 8.7% | 6.6% | 2.2% | | | |
| - veneer | 2.1 | 2.8 | 3.8 | 4.8 | 3.0% | 3.1% | 2.3% | | | |
| - plywood | 10.4 | 23.5 | 36.4 | 44.9 | 8.5% | 4.5% | 2.1% | | | |
| - particleboard | 0.5 | 8.9 | 32.4 | 45.3 | 34.1% | 13.8% | 3.4% | | | |
| - comp fibreboard | 1.7 | 4.3 | 7.9 | 9.3 | 9.4% | 6.4% | 1.6% | | | |
| - non-comp f'board | 5.9 | 7.6 | 8.7 | 6.9 | 2.6% | 1.4% | -2.2% | | | |
| Imports | 2.0 | 5.6 | 13.6 | 19.1 | 11.1% | 9.2% | 3.5% | | | |
| - veneer | 0.3 | 0.7 | 1.4 | 1.9 | 9.1% | 7.5% | 3.1% | | | |
| - plywood | 0.9 | 2.5 | 5.9 | 8.0 | 11.1% | 8.9% | 3.0% | | | |
| - particleboard | 0.1 | 0.9 | 4.2 | 6.9 | 27.4% | 17.1% | 5.1% | | | |
| - comp fibreboard | 0.4 | 1.0 | 1.3 | 1.4 | 8.0% | 2.9% | 0.7% | | | |
| - non-comp f'board | 0.3 | 0.6 | 0.7 | 1.0 | 8.2% | 2.1% | 3.1% | | | |
| Exports | 2.0 | 5.7 | 13.3 | 19.1 | 11.2% | 8.9% | 3.7% | | | |
| - veneer | 0.2 | 0.6 | 1.2 | 1.9 | 10.8% | 7.2% | 4.9% | | | |
| - plywood | 0.9 | 2.5 | 5.6 | 8.7 | 10.7% | 8.3% | 4.4% | | | |
| - particleboard | 0.1 | 0.9 | 4.4 | 6.3 | 27.8% | 17.2% | 3.6% | | | |
| - comp fibreboard | 0.5 | 1.0 | 1.3 | 1.4 | 7.7% | 3.2% | 0.8% | | | |
| - non-comp f'board | 0.3 | 0.7 | 0.8 | 0.9 | 8.8% | 1.4% | 1.3% | | | |
| Apparent Consumption | 20.5 | 47.0 | 89.5 | 111.2 | 8.7% | 6.7% | 2.2% | 1.80 | 1.60 | 0.76 |
| - veneer | 2.1 | 2.9 | 4.0 | 4.8 | 3.0% | 3.4% | 1.7% | 0.63 | 0.81 | 0.59 |
| - plywood | 10.3 | 23.5 | 36.7 | 44.2 | 8.6% | 4.6% | 1.9% | 1.79 | 1.10 | 0.65 |
| - particleboard | 0.5 | 8.9 | 32.3 | 45.9 | 34.1% | 13.8% | 3.6% | 7.10 | 3.32 | 1.24 |
| - comp fibreboard | 1.7 | 4.2 | 7.9 | 9.2 | 9.5% | 6.4% | 1.6% | 1.99 | 1.55 | 0.56 |
| - non-comp f'board | 5.9 | 7.5 | 8.6 | 7.1 | 2.5% | 1.4% | -2.0% | 0.52 | 0.34 | -0.69 |

GDP growth for elasticity calculations

| | 1955-65 | 1965-75 | 1975-85 |
|---|---|---|---|
| Developing countries | 4.9% | 5.6% | 4.4% |
| Industrial countries | 4.8% | 3.9% | 2.6% |
| World | 4.8% | 4.2% | 2.9% |

Table C.3 - Development of Paper and Paperboard Demand, 1955-1985

| | Amount (millions of tons 3-year averages) | | | | Average Annual Increase | | | Period Demand/Income Elasticity | | |
|---|---|---|---|---|---|---|---|---|---|---|
| | 1955 | 1965 | 1975 | 1985 | 1955-65 | 1965-75 | 1975-85 | 1955-65 | 1965-75 | 1975-85 |
| DEVELOPING COUNTRIES | | | | | | | | | | |
| Production | 1.4 | 5.7 | 12.4 | 27.1 | 15.4% | 8.0% | 8.2% | | | |
| - newsprint | 0.2 | 0.8 | 1.0 | 2.1 | 12.7% | 3.0% | 7.3% | | | |
| - printing/writing | 0.4 | 1.5 | 3.5 | 7.4 | 13.8% | 8.8% | 7.9% | | | |
| - other grades | 0.7 | 3.5 | 7.9 | 17.6 | 16.9% | 8.6% | 8.4% | | | |
| Imports | 1.5 | 2.5 | 4.5 | 7.1 | 5.1% | 5.9% | 4.5% | | | |
| - newsprint | 0.7 | 1.0 | 1.4 | 1.9 | 3.6% | 3.3% | 2.7% | | | |
| - printing/writing | 0.3 | 0.4 | 0.8 | 1.4 | 4.4% | 7.1% | 5.5% | | | |
| - other grades | 0.6 | 1.1 | 2.3 | 3.8 | 7.2% | 7.5% | 5.3% | | | |
| Exports | 0.1 | 0.2 | 0.5 | 1.6 | 9.7% | 9.7% | 12.7% | | | |
| - newsprint | .0 | 0.1 | 0.1 | 0.2 | 17.8% | 3.7% | 5.9% | | | |
| - printing/writing | .0 | .0 | 0.1 | 0.5 | 9.5% | 16.7% | 15.4% | | | |
| - other grades | 0.1 | 0.1 | 0.3 | 0.9 | 6.3% | 10.8% | 13.4% | | | |
| Apparent Consumption | 2.8 | 8.1 | 16.4 | 32.6 | 11.0% | 7.4% | 7.1% | 2.27 | 1.31 | 1.63 |
| - newsprint | 0.9 | 1.7 | 2.4 | 3.8 | 6.2% | 3.2% | 4.8% | 1.29 | 0.57 | 1.11 |
| - printing/writing | 0.7 | 1.9 | 4.1 | 8.3 | 10.9% | 8.3% | 7.2% | 2.26 | 1.48 | 1.66 |
| - other grades | 1.2 | 4.5 | 9.9 | 20.5 | 13.8% | 8.3% | 7.5% | 2.84 | 1.48 | 1.73 |
| INDUSTRIAL COUNTRIES | | | | | | | | | | |
| Production | 54.8 | 90.6 | 130.6 | 167.7 | 5.2% | 3.7% | 2.5% | | | |
| - newsprint | 11.0 | 16.0 | 20.9 | 26.4 | 3.8% | 2.7% | 2.4% | | | |
| - printing/writing | 9.9 | 17.6 | 28.2 | 44.2 | 6.0% | 4.8% | 4.6% | | | |
| - other grades | 33.9 | 57.0 | 81.5 | 97.1 | 5.3% | 3.6% | 1.8% | | | |
| Imports | 8.1 | 13.2 | 21.7 | 33.8 | 5.0% | 5.1% | 4.5% | | | |
| - newsprint | 5.9 | 7.5 | 9.4 | 12.1 | 2.4% | 2.3% | 2.5% | | | |
| - printing/writing | 0.3 | 1.2 | 3.7 | 8.8 | 13.6% | 11.8% | 9.1% | | | |
| - other grades | 1.8 | 4.5 | 8.6 | 12.9 | 9.4% | 6.7% | 4.2% | | | |
| Exports | 9.6 | 15.8 | 26.3 | 39.6 | 5.1% | 5.2% | 4.2% | | | |
| - newsprint | 6.6 | 8.5 | 10.3 | 13.4 | 2.6% | 1.9% | 2.7% | | | |
| - printing/writing | 0.6 | 1.7 | 5.0 | 9.7 | 10.2% | 11.5% | 6.8% | | | |
| - other grades | 2.4 | 5.5 | 11.0 | 16.5 | 8.9% | 7.1% | 4.2% | | | |
| Apparent Consumption | 53.3 | 88.0 | 126.0 | 161.9 | 5.1% | 3.7% | 2.5% | 1.08 | 0.94 | 0.99 |
| - newsprint | 10.3 | 14.9 | 20.1 | 25.1 | 3.8% | 3.0% | 2.2% | 0.80 | 0.77 | 0.88 |
| - printing/writing | 9.6 | 17.1 | 26.8 | 43.3 | 6.0% | 4.6% | 4.9% | 1.26 | 1.18 | 1.92 |
| - other grades | 33.4 | 55.9 | 79.1 | 93.5 | 5.3% | 3.5% | 1.7% | 1.11 | 0.90 | 0.66 |

Table C.3 (continued)

| | Amount (millions of tons 3-year averages) | | | | Average Annual Increase | | | Period Demand/Income Elasticity | | |
|---|---|---|---|---|---|---|---|---|---|---|
| | 1955 | 1965 | 1975 | 1985 | 1955-65 | 1965-75 | 1975-85 | 1955-65 | 1965-75 | 1975-85 |
| WORLD TOTALS | | | | | | | | | | |
| Production | 56.1 | 96.3 | 143.0 | 194.8 | 5.5% | 4.0% | 3.1% | | | |
| - newsprint | 11.2 | 16.7 | 21.9 | 28.5 | 4.1% | 2.7% | 2.6% | | | |
| - printing/writing | 10.3 | 19.1 | 31.7 | 51.7 | 6.4% | 5.2% | 5.0% | | | |
| - other grades | 34.7 | 60.4 | 89.4 | 114.7 | 5.7% | 4.0% | 2.5% | | | |
| Imports | 9.7 | 15.7 | 26.2 | 40.8 | 5.0% | 5.2% | 4.5% | | | |
| - newsprint | 6.7 | 8.5 | 10.9 | 14.0 | 2.5% | 2.5% | 2.5% | | | |
| - printing/writing | 0.6 | 1.6 | 4.5 | 10.1 | 10.4% | 10.8% | 8.6% | | | |
| - other grades | 2.4 | 5.6 | 10.9 | 16.7 | 8.9% | 6.9% | 4.4% | | | |
| Exports | 9.7 | 15.9 | 26.7 | 41.2 | 5.1% | 5.3% | 4.4% | | | |
| - newsprint | 6.6 | 8.6 | 10.4 | 13.6 | 2.7% | 1.9% | 2.7% | | | |
| - printing/writing | 0.6 | 1.7 | 5.1 | 10.1 | 10.2% | 11.6% | 7.0% | | | |
| - other grades | 2.4 | 5.6 | 11.3 | 17.5 | 8.9% | 7.2% | 4.5% | | | |
| Apparent Consumption | 56.1 | 96.1 | 142.4 | 194.5 | 5.5% | 4.0% | 3.2% | 1.15 | 0.97 | 1.09 |
| - newsprint | 11.2 | 16.7 | 22.4 | 28.9 | 4.0% | 3.0% | 2.5% | 0.84 | 0.73 | 0.88 |
| - printing/writing | 10.2 | 19.0 | 31.0 | 51.7 | 6.4% | 5.0% | 5.2% | 1.33 | 1.21 | 1.81 |
| - other grades | 34.6 | 60.4 | 89.0 | 114.0 | 5.7% | 4.0% | 2.5% | 1.19 | 0.95 | 0.86 |

GDP growth for elasticity calculations

| | 1955-65 | 1965-75 | 1975-85 |
|---|---|---|---|
| Developing countries | 4.9% | 5.6% | 4.4% |
| Industrial countries | 4.8% | 3.9% | 2.6% |
| World | 4.8% | 4.2% | 2.9% |

Table C.4 - Projected Demand for Forest Products, 1995 and 2000

| | Actual Consumption Average 1984-86 a/ | Actual Demand Elasticiy 1975-85 | Projected Demand Elasticity 1985-2000 b | | Projected Consumption | | | | | |
|---|---|---|---|---|---|---|---|---|---|---|
| | | | | | Low Growth c/ | | High Growth c/ | | Median Projection | |
| | | | Low | High | 1995 | 2000 | 1995 | 2000 | 1995 | 2000 |
| **DEVELOPING COUNTRIES** | | | | | | | | | | |
| Sawnwood (million cu m) | | | | | | | | | | |
| - hardwood | 59 | 1.1 | 1.0 | 1.2 | 72 | 80 | 94 | 118 | 83 | 99 |
| - softwood | 45 | 1.2 | 1.0 | 1.2 | 55 | 61 | 72 | 90 | 63 | 76 |
| Wood-based panels (million cu m) | | | | | | | | | | |
| - plywood & veneer | 10 | 2.3 | 1.5 | 1.9 | 14 | 16 | 20 | 29 | 17 | 22 |
| - particleboard | 3 | 2.1 | 1.6 | 2.0 | 5 | 6 | 7 | 11 | 6 | 8 |
| - fibreboard | 2 | 1.6 | 1.3 | 1.6 | 3 | 3 | 4 | 5 | 3 | 4 |
| Paper & paperboard (million tons) | | | | | | | | | | |
| - newsprint | 4 | 1.1 | 1.1 | 1.2 | 5 | 5 | 6 | 8 | 5 | 6 |
| - printing/writing | 8 | 1.7 | 1.4 | 1.6 | 11 | 13 | 15 | 21 | 13 | 17 |
| - other grades | 20 | 1.7 | 1.5 | 1.7 | 28 | 32 | 39 | 54 | 33 | 43 |
| **INDUSTRIAL COUNTRIES** | | | | | | | | | | |
| Sawnwood (million cu m) | | | | | | | | | | |
| - hardwood | 57 | -0.6 | -0.4 | 0.0 | 52 | 50 | 57 | 57 | 55 | 53 |
| - softwood | 302 | 0.2 | 0.0 | 0.1 | 302 | 302 | 313 | 319 | 307 | 310 |
| Wood-based panels (million cu m) | | | | | | | | | | |
| - plywood & veneer | 39 | 0.2 | 0.1 | 0.2 | 40 | 40 | 42 | 44 | 41 | 42 |
| - particleboard | 42 | 1.3 | 0.9 | 1.1 | 51 | 56 | 64 | 79 | 57 | 67 |
| - fibreboard | 14 | -0.3 | -0.4 | -0.3 | 13 | 12 | 13 | 12 | 13 | 12 |
| Paper & paperboard (million tons) | | | | | | | | | | |
| - newsprint | 25 | 0.9 | 0.7 | 0.8 | 29 | 31 | 34 | 39 | 31 | 35 |
| - printing/writing | 43 | 1.9 | 1.2 | 1.5 | 55 | 62 | 75 | 100 | 65 | 81 |
| - other grades | 93 | 0.7 | 0.4 | 0.6 | 101 | 105 | 117 | 131 | 109 | 118 |
| **WORLD** | | | | | | | | | | |
| Sawnwood (million cu m) | | | | | | | | | | |
| - hardwood | 116 | 0.4 | 0.4 | 0.7 | 125 | 130 | 151 | 175 | 138 | 153 |
| - softwood | 347 | 0.3 | 0.1 | 0.3 | 357 | 363 | 385 | 409 | 371 | 386 |
| - total sawnwood | 463 | 0.3 | 0.2 | 0.4 | 482 | 493 | 535 | 584 | 509 | 538 |
| Wood-based panels (million cu m) | | | | | | | | | | |
| - plywood & veneer | 49 | 0.6 | 0.5 | 0.7 | 53 | 56 | 63 | 73 | 58 | 65 |
| - particleboard | 46 | 1.2 | 1.0 | 1.2 | 56 | 61 | 71 | 89 | 63 | 75 |
| - fibreboard | 16 | .0 | -0.2 | 0.1 | 16 | 16 | 17 | 17 | 16 | 16 |
| - total panels | 111 | 0.8 | 0.6 | 0.8 | 125 | 133 | 151 | 180 | 138 | 156 |
| Paper & paperboard (million tons) | | | | | | | | | | |
| - newsprint | 29 | 0.9 | 0.8 | 0.8 | 34 | 36 | 40 | 47 | 37 | 41 |
| - printing/writing | 52 | 1.8 | 1.2 | 1.5 | 66 | 74 | 91 | 120 | 78 | 97 |
| - other grades | 114 | 0.9 | 0.6 | 0.8 | 129 | 137 | 156 | 185 | 142 | 161 |
| - total | 194 | 1.1 | 0.8 | 1.0 | 228 | 248 | 287 | 352 | 257 | 299 |

Notes:
(a) Annual average, 1985 to 1986, from 'Yearbook of Forest Products'. FAO. Rome, 1987.
(b) For basis of selection of projected demand elasticities, see text, Chapter II.
(c) Low growth economic scenario assumes 2.0% annual GDP growth through 2000. High growth scenario assumes 3.8% annually for industrial countries and 3.9% annually for developing countries over the same period.

## Table C.5 - Estimated Commercial Forest Area in Africa (1985)

| AFRICA | Natural Forests | | | | Industrial Plantations | | | Total Commercial Forests | Popul -ation | Forest Area per Capita | Removals (1985) | | Total Removal Rate |
|---|---|---|---|---|---|---|---|---|---|---|---|---|---|
| | Conif- erous | Broad- leaved | Bamboo | Total | Conif- erous | Broad- leaved | Total | | | | Fuelwood | Industry | |
| | ...................thousands of hectares | | | | | | | ................. | (million) | (ha/cap) | (thousand cu m) | | (cu m/ha) |
| Angola | - | 2230 | - | 2230 | 21 | 51 | 72 | 2302 | 9.9 | 0.23 | 7663 | 1433 | 4.0 |
| Benin | - | 41 | - | 41 | - | 8 | 8 | 49 | 3.9 | 0.01 | 4178 | 223 | 90.4 |
| Botswana | - | - | - | - | - | - | - | - | 1.0 | - | 729 | 74 | NA |
| Burkina Faso | - | - | - | - | - | - | - | - | 6.6 | - | 6386 | 305 | NA |
| Burundi | - | 5 | 12 | 17 | 14 | - | 14 | 31 | 4.6 | 0.01 | 3590 | 45 | NA |
| Cameroon | - | 16540 | - | 16540 | 1 | 13 | 15 | 16555 | 9.9 | 1.67 | 7994 | 2758 | 0.6 |
| C. Afr. R | - | 3445 | - | 3445 | - | - | - | 3445 | 2.5 | 1.38 | 2925 | 493 | 1.0 |
| Chad | - | 500 | - | 500 | - | - | - | 500 | 4.9 | 0.10 | 3062 | 505 | 7.1 |
| Congo, P.R. | - | 13580 | - | 13580 | 4 | 19 | 23 | 13603 | 1.8 | 7.56 | 1584 | 797 | 0.2 |
| Eq. Guinea | - | 1000 | - | 1000 | - | - | - | 1000 | 0.4 | - | 447 | 160 | 0.6 |
| Ethiopia | 400 | 510 | 800 | 1710 | 1 | - | 1 | 1711 | 42.2 | 0.04 | 36083 | 1813 | 22.1 |
| Gabon | - | 19830 | - | 19830 | - | 24 | 24 | 19854 | 0.8 | 24.45 | 1222 | 1484 | 0.1 |
| Gambia, The | - | 5 | - | 5 | - | 2 | 2 | 7 | 0.7 | 0.01 | 762 | 21 | NA |
| Ghana | - | 1211 | - | 1211 | 1 | 30 | 32 | 1243 | 12.3 | 0.10 | 7284 | 1175 | 6.8 |
| Guinea | - | 1270 | - | 1270 | 1 | 2 | 3 | 1273 | 5.9 | 0.22 | 3085 | 604 | 2.9 |
| Guinea-Bis. | - | 340 | - | 340 | - | 1 | 1 | 341 | 0.9 | 0.39 | 422 | 137 | 1.6 |
| Ivory Coast | - | 1845 | - | 1845 | - | 66 | 66 | 1911 | 9.9 | 0.19 | 7972 | 4514 | 6.5 |
| Kenya | 140 | 255 | 165 | 560 | 183 | 9 | 192 | 752 | 19.6 | 0.04 | 30854 | 1555 | 43.1 |
| Liberia | - | 1100 | - | 1100 | 9 | 47 | 56 | 1156 | 2.1 | 0.55 | 3810 | 452 | 3.7 |
| Madagascar | - | 5920 | - | 5920 | 287 | - | 287 | 6207 | 9.9 | 0.63 | 5455 | 807 | 1.0 |
| Malawi | - | 40 | - | 40 | 80 | 10 | 91 | 131 | 6.8 | 0.02 | 6204 | 384 | 50.5 |
| Mali | - | - | - | - | - | 2 | 2 | 2 | 7.3 | .00 | 4584 | 308 | NA |
| Mozambique | - | 400 | - | 400 | 20 | 6 | 26 | 426 | 13.4 | 0.03 | 14270 | 961 | 35.8 |
| Namibia | - | - | - | - | - | - | - | - | 1.1 | - | 2000 | 400 | NA |
| Niger | - | - | - | - | - | - | - | - | 6.2 | - | 3678 | 242 | NA |
| Nigeria | - | 1470 | - | 1470 | 4 | 266 | 270 | 1740 | 96.5 | 0.02 | 87587 | 7979 | 54.9 |
| Rwanda | - | 42 | 19 | 61 | 7 | 2 | 9 | 70 | 5.8 | 0.01 | 5239 | 222 | 78.0 |
| Senegal | - | 14 | - | 14 | - | 4 | 4 | 18 | 6.4 | .00 | 3559 | 547 | NA |
| Sierra Leone | - | 189 | - | 189 | - | 6 | 6 | 195 | 3.7 | 0.05 | 7634 | 140 | 39.8 |
| Somalia | - | 33 | - | 33 | - | - | - | 33 | 5.2 | - | 4368 | 67 | NA |
| South Africa | - | 300 | - | 300 | 625 | 625 | 1250 | 1550 | 31.6 | 0.05 | 7000 | 11944 | 12.2 |
| Sudan | - | 320 | - | 320 | 3 | 62 | 65 | 385 | 21.3 | 0.02 | 17684 | 1840 | 50.7 |
| Swaziland | - | 0 | - | 0 | 79 | 27 | 106 | 106 | 0.7 | 0.15 | 560 | 1663 | 21.0 |
| Tanzania | - | 780 | - | 780 | 65 | 11 | 77 | 857 | 21.5 | 0.04 | 44087 | 1453 | 53.2 |
| Togo | - | 243 | - | 243 | - | 6 | 6 | 248 | 2.9 | 0.09 | 602 | 163 | 3.1 |
| Uganda | - | 615 | - | 615 | 15 | 3 | 18 | 633 | 15.0 | 0.04 | 10856 | 1632 | 19.7 |
| Zaire | - | 79220 | 100 | 79320 | 1 | 18 | 20 | 79340 | 29.7 | 2.67 | 27971 | 2520 | 0.4 |
| Zambia | - | 2120 | - | 2120 | 30 | 14 | 44 | 2164 | 6.4 | 0.34 | 9418 | 473 | 4.6 |
| Zimbabwe | - | 500 | - | 500 | 197 | 68 | 265 | 765 | 8.1 | 0.09 | 5733 | 1376 | 9.3 |
| Totals | 540 | 155912 | 1096 | 157548 | 1650 | 1401 | 3051 | 160599 | | 0.29 | 398541 | 53669 | 2.8 |

Sources: (1) For South Africa: American University; "South Africa - A Country Study". Washington, DC. 1980.
(2) For Swaziland: Central Statistical Office; "Timber Statistics for 1982". Mbabane, 1984.
(3) For Zimbabwe: Central Statistical Office; "Agricultural Production in Rhodesia". Salisbury, 1969.
(4) For all other countries: Food and Agriculture Organization of the United Nations; "Les Ressources Forestieres de l'Afrique Tropicale". Rome, 1981.

Note: Source data is at various years and has been adjusted to account for estimated felling and new planting through 1985.

## Table C.6 - Estimated Commercial Forest Area in North and South America (1985)

| | Natural Forests | | | | Industrial Plantations | | | Total Commercial Forests | Population | Forest Area per Capita | Removals (1985) | | Total Removal Rate |
|---|---|---|---|---|---|---|---|---|---|---|---|---|---|
| | Coniferous | Broad-leaved | Bamboo | Total | Coniferous | Broad-leaved | Total | | | | Fuelwood | Industry | |
| | ....................thousands of hectares | | | | | | | .................. | (million) | (ha/cap) | (thousand cu m) | | (cu m/ha) |
| **NORTH-CENTRAL AMERICA** | | | | | | | | | | | | | |
| Bahamas | 159 | - | - | 159 | - | - | - | 159 | 0.2 | 0.69 | - | 115 | 0.7 |
| Belize | 90 | 776 | - | 866 | 2 | 1 | 3 | 869 | 0.2 | 5.57 | 126 | 38 | 0.2 |
| Canada | 148000 | 67000 | - | 215000 | - | - | - | 215000 | 25.1 | 8.57 | 6197 | 165108 | 0.8 |
| Costa Rica | - | 860 | - | 860 | 2 | 3 | 5 | 865 | 2.5 | 0.35 | 2551 | 504 | 3.5 |
| Cuba | 200 | 790 | - | 990 | 105 | 109 | 215 | 1205 | 9.9 | 0.12 | 2819 | 525 | 2.8 |
| Dominican R. | 176 | 386 | - | 562 | - | - | - | 562 | 6.1 | 0.09 | 976 | 6 | 1.7 |
| El Salvador | 20 | 3 | - | 23 | 1 | 2 | 2 | 25 | 5.4 | .00 | 4670 | 84 | NA |
| Guatemala | 612 | 2175 | - | 2787 | 34 | 22 | 56 | 2843 | 7.7 | 0.37 | 6674 | 195 | 2.4 |
| Haiti | 11 | 10 | - | 21 | - | - | - | 21 | 5.4 | .00 | 5663 | 239 | NA |
| Honduras | 1554 | 1309 | - | 2863 | - | - | - | 2863 | 4.2 | 0.68 | 4538 | 544 | 1.8 |
| Jamaica | | 43 | - | 43 | 11 | 6 | 16 | 59 | 2.2 | 0.03 | 13 | 80 | 1.6 |
| Mexico | 11720 | 11355 | - | 23075 | 63 | 59 | 122 | 23197 | 76.8 | 0.30 | 13844 | 7473 | 0.9 |
| Nicaragua | 326 | 3225 | - | 3551 | 6 | - | 6 | 3557 | 3.2 | 1.11 | 2701 | 880 | 1.0 |
| Panama | - | 2808 | - | 2808 | 5 | 1 | 6 | 2814 | 2.1 | 1.34 | 1708 | 339 | 0.7 |
| Trin. & Tob. | - | 145 | - | 145 | 7 | 12 | 19 | 164 | 1.2 | 0.14 | 22 | 41 | 0.4 |
| USA | 84000 | 111000 | - | 195000 | - | - | - | 195000 | 240.7 | 0.81 | 101922 | 346566 | 2.3 |
| Totals | 246868 | 201884 | - | 448752 | 235 | 215 | 449 | 449201 | | 1.39 | 154424 | 522737 | 1.5 |
| **SOUTH AMERICA** | | | | | | | | | | | | | |
| Argentina | - | 38600 | - | 38600 | 428 | 122 | 550 | 39150 | 30.1 | 1.30 | 8462 | 4967 | 0.3 |
| Bolivia | - | 29475 | - | 29475 | 1 | 7 | 8 | 29483 | 6.2 | 4.76 | 1168 | 149 | .0 |
| Brazil | 280 | 295480 | - | 295760 | 1902 | 1144 | 3046 | 298806 | 132.6 | 2.25 | 168124 | 57781 | 0.8 |
| Chile | - | 7616 | - | 7616 | 1112 | 115 | 1227 | 8843 | 11.8 | 0.75 | 6076 | 9417 | 1.8 |
| Colombia | - | 36000 | - | 36000 | 88 | 50 | 138 | 36138 | 28.4 | 1.27 | 14551 | 2673 | 0.5 |
| Ecuador | - | 9685 | - | 9685 | 13 | 52 | 66 | 9751 | 9.1 | 1.07 | 5879 | 2692 | 0.9 |
| Fr. Guiana | - | 7590 | - | 7590 | - | - | - | 7590 | 0.1 | 115.00 | 66 | 188 | .0 |
| Guyana | - | 13455 | - | 13455 | 1 | - | 1 | 13456 | 0.8 | 17.14 | 12 | 180 | .0 |
| Paraguay | - | 2190 | - | 2190 | 3 | - | 3 | 2193 | 3.3 | 0.66 | 4813 | 2747 | 3.4 |
| Peru | 180 | 42795 | - | 42975 | - | 47 | 47 | 43022 | 18.2 | 2.36 | 6513 | 1210 | 0.2 |
| Suriname | - | 12485 | - | 12485 | 9 | - | 9 | 12494 | 0.4 | 32.62 | 8 | 209 | .0 |
| Uruguay | - | 185 | - | 185 | - | - | - | 185 | 3.0 | 0.06 | 2762 | 213 | 16.1 |
| Venezuela | - | 18835 | - | 18835 | 196 | 15 | 211 | 19046 | 16.8 | 1.13 | 666 | 636 | 0.1 |
| Totals | 460 | 514391 | - | 514851 | 3752 | 1552 | 5304 | 520155 | | 2.04 | 219100 | 83062 | 0.6 |

Sources: (1) For Canada and the USA: United Nations Economic Commission for Europe and the Food and Agriculture Organization of the United Nations; "The Forest Resources of the ECE Region". Geneva, 1985.
(2) For Bahamas: P.T.Henry; "The Pine Forests of the Bahamas". ODA, London, 1974.
(3) For Argentina, Chile and Uruguay: World Bank data.
(4) For all other countries: Food and Agriculture Organization of the United Nations; "Los Recursos Forestales de la America Tropical". Rome, 1981.

Note: Source data is at various years and has been adjusted to account for estimated felling and new planting through 1985.

## Table C.7 - Estimated Commercial Forest Area in Asia and Oceania (1985)

| | Natural Forests | | | | Industrial Plantations | | | Total Commercial Forests | Population | Forest Area per Capita | Removals (1985) | | Total Removal Rate |
|---|---|---|---|---|---|---|---|---|---|---|---|---|---|
| | Conif-erous | Broad-leaved | Bamboo | Total | Conif-erous | Broad-leaved | Total | | | | Fuelwood | Industry | |
| | ...thousands of hectares... | | | | | | | | (million) | (ha/cap) | (thousand cu m) | | (cu m/ha) |
| **ASIA** | | | | | | | | | | | | | |
| Bangladesh | - | 810 | - | 810 | - | 213 | 213 | 1023 | 98.1 | 0.01 | 26298 | 846 | 26.5 |
| Bhutan | 510 | 1308 | - | 1818 | 3 | 8 | 11 | 1829 | 1.2 | 1.52 | 2946 | 278 | 1.8 |
| Brunei | - | 262 | - | 262 | - | - | - | 262 | 0.2 | 1.20 | 79 | 215 | 1.1 |
| Burma | 116 | 22591 | 632 | 23339 | - | 2 | 2 | 23341 | 36.1 | 0.65 | 15853 | 3023 | 0.8 |
| China | 42000 | 34000 | 4000 | 80000 | 16800 | 13200 | 30000 | 110000 | 1029.2 | 0.11 | 170721 | 92652 | 2.4 |
| Cyprus | 70 | 70 | - | 140 | 28 | - | 28 | 168 | 0.7 | 0.26 | 23 | 57 | 0.5 |
| India | 3147 | 37623 | 1440 | 42210 | 76 | 1940 | 2016 | 44226 | 749.2 | 0.06 | 222437 | 22592 | 5.5 |
| Indonesia | 160 | 70575 | - | 70735 | 509 | 1202 | 1711 | 72446 | 158.9 | 0.46 | 122197 | 26811 | 2.1 |
| Iran, I.R. | 500 | 2600 | - | 3100 | - | - | - | 3100 | 43.8 | 0.07 | 2369 | 4376 | 2.2 |
| Japan | 2500 | 11677 | 141 | 14318 | 9435 | 149 | 9584 | 23902 | 120.0 | 0.20 | 521 | 32944 | 1.4 |
| Kampuchea | 8 | 4995 | 380 | 5383 | - | 6 | 6 | 5389 | 5.0 | 1.08 | 4736 | 567 | 1.0 |
| Korea, D.P.R | 500 | 1000 | - | 1500 | - | - | - | 1500 | 19.9 | 0.08 | 3943 | 600 | 3.0 |
| Korea, R. of | 1000 | 500 | - | 1500 | 267 | 133 | 400 | 1900 | 40.1 | 0.05 | 6178 | 2395 | 4.5 |
| Lao PDR | 100 | 2380 | 600 | 3080 | - | 6 | 6 | 3086 | 3.5 | 0.88 | 3807 | 244 | 1.3 |
| Malaysia | - | 14277 | - | 14277 | 53 | 73 | 127 | 14404 | 15.3 | 0.94 | 7582 | 32106 | 2.8 |
| Mongolia | 5000 | 200 | - | 5200 | - | - | - | 5200 | 1.9 | 2.74 | 1350 | 1040 | 0.5 |
| Nepal | 215 | 635 | 1 | 851 | 15 | 5 | 20 | 871 | 16.1 | 0.05 | 15216 | 560 | 18.1 |
| Pakistan | 835 | 185 | - | 1020 | - | - | - | 1020 | 92.4 | 0.01 | 19613 | 620 | 19.8 |
| Philippines | 190 | 6245 | - | 6435 | 12 | 103 | 115 | 6550 | 53.4 | 0.12 | 29172 | 7442 | 5.6 |
| Sri Lanka | - | 936 | - | 936 | 8 | 159 | 167 | 1103 | 15.9 | 0.07 | 7873 | 703 | 7.8 |
| Thailand | 165 | 2655 | 900 | 3720 | 4 | 111 | 115 | 3835 | 50.0 | 0.08 | 37053 | 4223 | 10.8 |
| Turkey | 3874 | 2768 | - | 6642 | - | - | - | 6642 | 48.4 | 0.14 | 12488 | 5958 | 2.8 |
| Viet Nam | 110 | 3345 | 1200 | 4655 | 67 | 10 | 77 | 4732 | 60.1 | 0.08 | 21622 | 3250 | 5.3 |
| Totals | 61000 | 221637 | 9294 | 291931 | 17941 | 17172 | 44597 | 336528 | | 0.12 | 734077 | 243502 | 2.9 |
| **OCEANIA** | | | | | | | | | | | | | |
| Australia | - | 40938 | - | 40938 | 793 | 45 | 838 | 41776 | 15.5 | 2.70 | 2874 | 16343 | 0.5 |
| Fiji | - | 310 | - | 310 | 64 | - | 64 | 374 | 0.7 | 0.55 | 37 | 212 | 0.7 |
| New Zealand | - | 1000 | - | 1000 | 1045 | 44 | 1089 | 2089 | 3.2 | 0.65 | 50 | 8884 | 4.3 |
| P. N. Guinea | 50 | 13925 | - | 13975 | 17 | 5 | 22 | 13997 | 3.4 | 4.12 | 5533 | 2090 | 0.5 |
| Totals | 50 | 56173 | - | 56223 | 1919 | 94 | 2013 | 58236 | | 2.36 | 8494 | 27529 | 0.6 |

Sources: (1) For China: World Bank data.
(2) For Iran, Democratic Peoples' Republic of Korea, and the Republic of Korea: American University Country Studies. Washington, DC. 1978
(3) For Japan: Ministry of Agriculture, Forestry & Fisheries; "Abstract of Statistics on Agriculture Forestry and Fisheries". Tokyo, 1982.
(4) For Mongolia: Crosfield (Joseph) & Sons Ltd.; "Mongolia - An Economic Handbook". London, 1965.
(5) For Australia: Australia Bureau of Statistics; "Year Book Australia 1986". Canberra, 1986.
(6) For Fiji: Ministry of Forests; "Timber Prpfile Fiji". Suva, 1985.
(7) For New Zealand: New Zealand Department of Statistics; "New Zealand Official Yearbook". Wellington, 1985.
(8) For all other countries: Food and Agriculture Organization of the United Nations; "Forest Resources of Tropical Asia". Rome, 1981.

Note: Source data is at various years and has been adjusted to account for estimated felling and new planting through 1985.

Table C.8 - Estimated Commercial Forest Area in Europe and the USSR and World Totals (1985)

| | Natural Forests | | | | Industrial Plantations | | | | | | Removals (1985) | | |
|---|---|---|---|---|---|---|---|---|---|---|---|---|---|
| | Conif-erous | Broad-leaved | Bamboo | Total | Conif-erous | Broad-leaved | Total | Total Commercial Forests | Popul-ation | Forest Area per Capita | Fuelwood | Industry | Total Removal Rate |
| EUROPE | ....................thousands of hectares | | | | | | | .................. | (million) | (ha/cap) | (thousand cu m) | | (cu m/ha) |
| Albania | 100 | 830 | - | 930 | - | - | - | 930 | 2.9 | 0.32 | 1608 | 722 | 2.5 |
| Austria | 2503 | 662 | - | 3165 | - | - | - | 3165 | 7.6 | 0.42 | 1413 | 12791 | 4.5 |
| Belgium/Lux | 310 | 370 | - | 680 | - | - | - | 680 | 10.3 | 0.07 | 538 | 2548 | 4.5 |
| Bulgaria | 1100 | 2200 | - | 3300 | - | - | - | 3300 | 9.0 | 0.37 | 1721 | 3065 | 1.5 |
| Czechoslovak | 2850 | 1585 | - | 4435 | - | - | - | 4435 | 15.5 | 0.29 | 1382 | 17620 | 4.3 |
| Denmark | 290 | 110 | - | 400 | - | - | - | 400 | 5.2 | 0.08 | 351 | 2342 | 6.7 |
| Finland | 17884 | 1561 | - | 19445 | - | - | - | 19445 | 4.9 | 3.97 | 3075 | 38707 | 2.1 |
| France | 4405 | 8935 | - | 13340 | - | - | - | 13340 | 54.9 | 0.24 | 10424 | 28575 | 2.9 |
| German Dem R | 1995 | 595 | - | 2590 | - | - | - | 2590 | 16.7 | 0.16 | 656 | 9910 | 4.1 |
| Germany, F.R | 4592 | 2246 | - | 6838 | - | - | - | 6838 | 61.2 | 0.11 | 3795 | 26855 | 4.5 |
| Greece | 690 | 1103 | - | 1793 | - | - | - | 1793 | 9.9 | 0.18 | 1915 | 768 | 1.5 |
| Hungary | 220 | 1343 | - | 1563 | - | - | - | 1563 | 10.7 | 0.15 | 3000 | 3780 | 4.3 |
| Ireland | 298 | 49 | - | 347 | - | - | - | 347 | 3.5 | 0.10 | 46 | 1210 | 3.6 |
| Israel | 41 | 25 | - | 66 | - | - | - | 66 | 4.2 | 0.02 | 11 | 107 | 1.8 |
| Italy | 976 | 2892 | - | 3868 | - | - | - | 3868 | 57.0 | 0.07 | 4891 | 4557 | 2.4 |
| Netherlands | 191 | 103 | - | 294 | - | - | - | 294 | 14.4 | 0.02 | 100 | 1039 | 3.9 |
| Norway | 5280 | 1320 | - | 6600 | - | - | - | 6600 | 4.1 | 1.61 | 799 | 9821 | 1.6 |
| Poland | 6681 | 1729 | - | 8410 | - | - | - | 8410 | 36.9 | 0.23 | 3510 | 19830 | 2.8 |
| Portugal | 1347 | 1243 | - | 2590 | - | - | - | 2590 | 10.2 | 0.25 | 500 | 8758 | 3.6 |
| Romania | 1580 | 4280 | - | 5860 | - | - | - | 5860 | 22.7 | 0.26 | 4569 | 19557 | 4.1 |
| Spain | 4823 | 1683 | - | 6506 | - | - | - | 6506 | 38.7 | 0.17 | 1863 | 11833 | 2.1 |
| Sweden | 21103 | 1127 | - | 22230 | - | - | - | 22230 | 8.3 | 2.68 | 4424 | 48915 | 2.4 |
| Switzerland | 550 | 245 | - | 795 | - | - | - | 795 | 6.4 | 0.12 | 870 | 4095 | 6.2 |
| Un. Kingdom | 1407 | 610 | - | 2017 | - | - | - | 2017 | 56.6 | 0.04 | 159 | 4269 | 2.2 |
| Yugoslavia | 1870 | 6630 | - | 8500 | - | - | - | 8500 | 23.0 | 0.37 | 4103 | 11760 | 1.9 |
| Totals | 83086 | 43476 | - | 126562 | - | - | - | 126562 | | 0.28 | 55723 | 293434 | 2.8 |
| USSR | 405900 | 128600 | - | 534500 | - | - | - | 534500 | 275.0 | 1.42 | 80300 | 275400 | 0.7 |
| World Total | 797904 | 1322073 | 10390 | 2130367 | 25497 | 20434 | 55415 | 2185782 | | 0.46 | 1650659 | 1499333 | 1.4 |

Source: United Nations Economic Commission for Europe and the Food and Agriculture Organization of the United Nations; "The Forest Resources of the ECE Region". Geneva, 1985.

Notes: (1) For the forests of Europe, which have been managed over many years, distinctions between "natural" and "planted" forests are difficult. In the above tabulation, therefore, all productive closed forest areas are included under "natural" forests.

(2) Source data is at various years and has been adjusted to account for estimated felling and new planting through 1985.

Table C.9 - Actual and Projected Industrial Roundwood Balance

| | Actual (1) | | | | Low Projection | | High Projection | | Median Projection | |
|---|---|---|---|---|---|---|---|---|---|---|
| | 1955 | 1965 | 1975 | 1985 | 1995 | 2000 | 1995 | 2000 | 1995 | 2000 |
| Industrial Log Supply (million cubic meters) | | | | | | | | | | |
| - Sawlogs & veneer logs | 578 | 681 | 792 | 898 | 939 | 963 | 1049 | 1153 | 994 | 1058 |
| - Pulplogs | 181 | 242 | 335 | 387 | 467 | 512 | 624 | 811 | 539 | 660 |
| - Other Industrial wood (2) | 160 | 184 | 205 | 235 | 230 | 230 | 230 | 230 | 230 | 230 |
| - Total | 919 | 1108 | 1332 | 1520 | 1636 | 1705 | 1903 | 2194 | 1763 | 1948 |
| Sawnwood, Plywood & Veneer Production (million cubic meters) | | | | | | | | | | |
| - Sawnwood | 309 | 373 | 422 | 464 | 482 | 493 | 535 | 584 | 509 | 538 |
| - Plywood | 10 | 24 | 36 | 45 | 53 | 56 | 63 | 73 | 58 | 65 |
| - Veneer | 2 | 3 | 4 | 5 | Incl | Incl | Incl | Incl | Incl | Incl |
| - Total | 321 | 400 | 462 | 514 | 535 | 549 | 598 | 657 | 567 | 603 |
| Apparent sawnwood/plywood yield (3) | 55.6% | 58.6% | 58.4% | 57.2% | 57.0% | 57.0% | 57.0% | 57.0% | 57.0% | 57.0% |
| Reconstituted Wood Panel Production (million cubic meters) | | | | | | | | | | |
| - Particleboard | 0 | 9 | 32 | 45 | 56 | 51 | 65 | 80 | 57 | 66 |
| - Compressed fibreboard | 2 | 4 | 8 | 9 | 10 | 10 | 11 | 11 | 10 | 10 |
| - Non-compressed fibreboard | 6 | 8 | 9 | 7 | 6 | 6 | 6 | 6 | 6 | 6 |
| - Total | 8 | 21 | 49 | 62 | 71 | 67 | 82 | 97 | 73 | 82 |
| Wood Pulp Production (million tons) | | | | | | | | | | |
| - Mechanical | 15 | 20 | 24 | 30 | 35 | 38 | 42 | 49 | 39 | 43 |
| - Semi-chemical | 4 | 6 | 8 | 7 | 7 | 7 | 7 | 7 | 7 | 7 |
| - Chemical | 27 | 47 | 74 | 93 | 111 | 122 | 144 | 180 | 127 | 151 |
| - Total | 46 | 72 | 105 | 131 | 153 | 166 | 192 | 236 | 172 | 201 |
| Pulpwood Balance (million cubic meters) | | | | | | | | | | |
| - Requirement (4) | | | | | | | | | | |
| - for panels | 7 | 23 | 62 | 80 | 94 | 87 | 108 | 129 | 96 | 109 |
| - for pulp | 182 | 298 | 446 | 556 | 655 | 713 | 831 | 1028 | 741 | 869 |
| - total requirement | 189 | 321 | 507 | 636 | 749 | 801 | 939 | 1157 | 837 | 978 |
| - Supply | | | | | | | | | | |
| - from pulplogs | 181 | 242 | 335 | 387 | 467 | 512 | 624 | 811 | 539 | 660 |
| - from sawmill/plywood mill waste | 8 | 79 | 172 | 249 | 282 | 289 | 315 | 346 | 298 | 317 |
| - total supply | 189 | 321 | 507 | 636 | 749 | 801 | 939 | 1157 | 837 | 978 |
| Apparent yield of waste from sawlogs and veneer logs (5) | 1.4% | 11.6% | 21.8% | 27.7% | 30.0% | 30.0% | 30.0% | 30.0% | 30.0% | 30.0% |

Notes: (1) Three-year averages

(2) The consumption of "other" industrial roundwood has been projected to continue at approximately the present level of 235 million cubic meters annually.

(3) The apparent yield of sawnwood, plywood and veneer is 57%. This same figure has been used to project future log requirements for these products

(4) Assumed roundwood equivalents are :

| | cu m/cu m | | cu m/ton |
|---|---|---|---|
| particleboard | 1.30 | mechanical pulp | 2.50 |
| compressed fibreboard | 1.90 | semi-chem pulp | 3.30 |
| non-compressed fibreboa | 0.50 | chemical pulp | 4.90 |

(5) The use of residues from sawmills and plywood mills as the raw material for pulpmills and reconsituted panel mills has increased from almost zero in 1955 to 28% of log input in 1985. The figure of 30% has been used to estimate future pulpwood from this source.

## Table C.10 - The Structure of Trade in Forest Products, Africa

| AFRICA | 1985 Trade in Forest Products | | | Forest Products in Total Trade | | Structure of 1985 Imports | | | Structure of 1985 Exports | | |
|---|---|---|---|---|---|---|---|---|---|---|---|
| | Imports | Exports | Net Exports | Imports | Exports | Logs | Mech Wood Products | Pulp & Paper | Logs | Mech Wood Products | Pulp & Paper |
| | .... (US$ million) ..... | | | .. (percent) .. | | ...... (percent) ....... | | | ...... (percent) ....... | | |
| Algeria | 299 | - | (299) | 3.3 | - | 12 | 63 | 26 | - | - | - |
| Angola | 0 | - | (0) | .0 | - | - | - | 100 | - | - | - |
| Benin | 6 | - | (6) | 1.3 | - | - | 15 | 85 | - | - | - |
| Botswana | 2 | - | (2) | NA | - | - | 66 | 34 | - | - | - |
| Burkina Faso | 2 | - | (2) | 0.9 | - | - | 64 | 36 | - | - | - |
| Burundi | 1 | - | (1) | 0.7 | - | - | 32 | 68 | - | - | - |
| Cameroon | 5 | 91 | 86 | 0.5 | 3.9 | - | - | 100 | 70 | 30 | - |
| Central Afr.Rep. | 2 | 20 | 19 | 1.2 | 17.6 | - | - | 100 | 58 | 42 | - |
| Chad | 1 | - | (1) | 0.6 | - | - | 78 | 22 | - | - | - |
| Congo, Peop.Rep. | 3 | 49 | 46 | 0.4 | 4.4 | - | 76 | 24 | 59 | 41 | - |
| Egypt, Arab Rep | 668 | - | (668) | 6.0 | - | 10 | 63 | 27 | - | - | - |
| Ethiopia | 10 | - | (10) | 1.0 | - | - | 2 | 98 | - | - | - |
| Gabon | 4 | 108 | 104 | 0.5 | 5.7 | - | - | 100 | 85 | 15 | - |
| Gambia, The | 0 | - | (0) | .0 | 0.0 | - | 100 | - | - | - | |
| Ghana | 3 | 27 | 24 | 0.4 | 4.3 | - | - | 100 | 36 | 64 | - |
| Ivory Coast | 22 | 228 | 206 | 1.3 | 7.7 | - | - | 100 | 57 | 43 | - |
| Kenya | 18 | 3 | (15) | 1.3 | 0.3 | - | 4 | 96 | 33 | 10 | 57 |
| Liberia | 1 | 36 | 34 | 0.5 | 7.9 | - | 69 | 31 | 95 | 5 | - |
| Libya | 111 | - | (111) | 1.8 | 0.0 | 9 | 75 | 16 | - | - | - |
| Madagascar | 7 | 0 | (7) | 2.2 | .0 | - | 0 | 100 | 100 | 0 | - |
| Malawi | 17 | 1 | (16) | 5.8 | 0.2 | - | 12 | 88 | - | 100 | - |
| Mali | 1 | - | (1) | 0.3 | - | - | 71 | 29 | - | - | - |
| Mauritius | 9 | - | (9) | 1.7 | - | 3 | 48 | 50 | - | - | - |
| Morocco | 117 | 24 | (93) | 3.0 | 1.1 | 19 | 48 | 33 | - | 15 | 85 |
| Mozambique | 0 | 1 | 1 | 0.1 | 0.7 | - | - | 100 | 49 | 51 | - |
| Niger | 0 | - | (0) | 0.1 | - | - | 39 | 61 | - | - | - |
| Nigeria | 160 | 6 | (154) | 1.8 | .0 | 0 | 6 | 94 | 98 | 2 | - |
| Rwanda | 2 | - | (2) | 0.8 | - | - | 51 | 49 | - | - | - |
| Senegal | 35 | - | (35) | 4.0 | - | 9 | 15 | 76 | - | - | - |
| Sierra Leone | 1 | - | (1) | 0.7 | - | - | 86 | 14 | - | - | - |
| Somalia | 2 | - | (2) | 0.6 | - | 4 | 76 | 20 | - | - | - |
| South Africa | 240 | 238 | (2) | 2.1 | 1.4 | 2 | 29 | 69 | 2 | 4 | 94 |
| Sudan | 22 | - | (22) | 2.9 | - | - | 41 | 59 | - | - | - |
| Swaziland | 1 | 83 | 83 | 0.1 | 11.9 | 100 | - | 0 | 5 | 13 | 83 |
| Tanzania | 18 | 1 | (17) | 1.8 | 0.3 | - | - | 100 | 7 | 93 | - |
| Togo | 0 | - | (0) | 0.2 | - | 18 | 6 | 76 | - | - | - |
| Tunisia | 117 | - | (117) | 4.3 | - | 3 | 56 | 41 | - | - | - |
| Uganda | 3 | 0 | (3) | 0.9 | .0 | - | 8 | 92 | - | 100 | - |
| Zaire | 6 | 18 | 12 | 0.5 | 1.1 | - | 13 | 87 | 37 | 63 | - |
| Zambia | 22 | - | (22) | 3.3 | - | 5 | 11 | 84 | - | - | - |
| Zimbabwe | 16 | 9 | (7) | 1.9 | 0.8 | 1 | 14 | 84 | 6 | 49 | 45 |
| Totals | 1,956 | 947 | (1,009) | 2.6 | 1.1 | 8 | 48 | 45 | 42 | 24 | 34 |

Source: IMF; "International Financial Statistics". Washington, DC. 1987.
FAO; "Yearbook of Forest Products Statistics". Rome. 1986.

## Table C.11 - The Structure of Trade in Forest Products, North and South America

| | 1985 Trade in Forest Products | | | Forest Products in Total Trade | | Structure of 1985 Imports | | | Structure of 1985 Exports | | |
|---|---|---|---|---|---|---|---|---|---|---|---|
| | Imports | Exports | Net Exports | Imports | Exports | Logs | Mech Wood Products | Pulp & Paper | Logs | Mech Wood Products | Pulp & Paper |
| | .... (US$ million) ..... | | | .. (percent) .. | | ...... (percent) ....... | | | ...... (percent) ....... | | |
| **NORTH-CENTRAL AMERICA** | | | | | | | | | | | |
| Bahamas | 7 | - | (7) | 0.2 | - | - | 93 | 7 | - | - | - |
| Barbados | 13 | - | (13) | 1.9 | - | 1 | 71 | 28 | - | - | - |
| Belize | 2 | 1 | (1) | 1.6 | 0.9 | - | 29 | 71 | 7 | 93 | - |
| Canada | 980 | 11,221 | 10,241 | 1.2 | 12.8 | 14 | 34 | 53 | 2 | 33 | 65 |
| Costa Rica | 67 | 17 | (50) | 6.0 | 1.8 | 0 | 0 | 99 | 1 | 43 | 56 |
| Cuba | 253 | - | (253) | NA | - | 1 | 58 | 41 | - | - | - |
| Dominica | 1 | - | (1) | 1.9 | - | 3 | 97 | - | - | - | - |
| Dominican Rep. | 49 | - | (49) | 3.8 | - | 7 | 19 | 74 | - | - | - |
| El Salvador | 37 | 7 | (30) | 3.7 | 0.9 | - | 15 | 85 | - | - | 100 |
| Guatemala | 45 | 11 | (33) | 3.8 | 1.1 | 0 | 2 | 98 | 1 | 26 | 73 |
| Haiti | 5 | - | (5) | 1.0 | - | - | 38 | 62 | - | - | - |
| Honduras | 21 | 31 | 10 | 3.6 | 7.7 | - | - | 100 | 2 | 98 | - |
| Jamaica | 49 | 1 | (48) | 4.4 | 0.3 | - | 29 | 71 | - | - | 100 |
| Mexico | 307 | 14 | (293) | 2.3 | 0.1 | 1 | 16 | 83 | 1 | 78 | 21 |
| Neth. Antilles | 10 | - | (10) | 0.2 | - | 4 | 55 | 41 | - | - | - |
| Nicaragua | 12 | 3 | (10) | 1.5 | 0.8 | 1 | 0 | 99 | - | 100 | - |
| Panama | 37 | 1 | (36) | 1.4 | 0.1 | 1 | 3 | 96 | 22 | 57 | 22 |
| Saint Vincent | 3 | - | (3) | 4.9 | - | - | 38 | 62 | - | - | - |
| Trin & Tobago | 101 | - | (101) | 6.3 | 0.0 | 1 | 65 | 34 | - | - | - |
| United States | 10,551 | 5,335 | (5,216) | 2.9 | 2.5 | 1 | 35 | 64 | 27 | 18 | 54 |
| Totals | 12,549 | 16,642 | 4,093 | 2.6 | 4.9 | 2 | 35 | 63 | 10 | 29 | 61 |
| **SOUTH AMERICA** | | | | | | | | | | | |
| Argentina | 95 | 9 | (85) | 2.5 | 0.1 | 2 | 49 | 49 | 7 | 37 | 57 |
| Bolivia | 11 | 6 | (5) | 2.0 | 0.9 | - | - | 100 | - | 100 | - |
| Brazil | 148 | 801 | 652 | 1.0 | 3.1 | 5 | 13 | 82 | 0 | 32 | 68 |
| Chile | 42 | 350 | 308 | 1.5 | 9.3 | - | - | 100 | 12 | 19 | 69 |
| Colombia | 117 | 14 | (103) | 2.8 | 0.4 | - | 6 | 94 | - | 8 | 92 |
| Ecuador | 121 | 20 | (102) | 7.6 | 0.7 | - | - | 100 | - | 100 | - |
| Guyana | 6 | 4 | (2) | 2.6 | 2.3 | - | 5 | 95 | 33 | 67 | - |
| Paraguay | 9 | 91 | 82 | 1.7 | 29.9 | - | - | 100 | - | 100 | - |
| Peru | 38 | 4 | (34) | 2.0 | 0.1 | 0 | 6 | 94 | - | 45 | 55 |
| Suriname | 10 | 4 | (5) | 2.8 | 1.2 | - | - | 100 | 39 | 61 | - |
| Uruguay | 14 | 7 | (7) | 2.1 | 0.8 | 0 | 24 | 75 | - | - | 100 |
| Venezuela | 287 | - | (287) | 3.5 | 0.0 | 5 | 21 | 74 | - | - | - |
| Totals | 896 | 1,309 | 413 | 2.3 | 2.1 | 3 | 15 | 82 | 4 | 35 | 62 |

Table C.12 - The Structure of Trade in Forest Products, Asia

| | 1985 Trade in Forest Products | | | Forest Products in Total Trade | | Structure of 1985 Imports | | | Structure of 1985 Exports | | |
|---|---|---|---|---|---|---|---|---|---|---|---|
| | Imports | Exports | Net Exports | Imports | Exports | Logs | Mech Wood Products | Pulp & Paper | Logs | Mech Wood Products | Pulp & Paper |
| | .... (US$ million) ..... | | | .. (percent) .. | | ...... (percent) ....... | | | ...... (percent) ....... | | |
| ASIA | | | | | | | | | | | |
| Afghanistan | 29 | - | (29) | 2.9 | 0.0 | - | 94 | 6 | - | - | - |
| Bahrain | 46 | - | (46) | NA | NA | 32 | 54 | 15 | - | - | - |
| Bangladesh | 10 | 9 | (1) | 0.4 | 1.0 | - | 2 | 98 | - | - | 100 |
| Bhutan | - | 1 | 1 | NA | NA | - | - | - | 43 | 57 | - |
| Brunei | 7 | 0 | (7) | NA | NA | - | 58 | 42 | - | 100 | - |
| Burma | 12 | 85 | 73 | 4.2 | 28.2 | - | - | 100 | 66 | 34 | - |
| China | 1,894 | 440 | (1,454) | 4.5 | 1.6 | 53 | 13 | 34 | 2 | 62 | 36 |
| Cyprus | 48 | - | (48) | NA | NA | 5 | 36 | 59 | - | - | - |
| Hong Kong | 540 | 78 | (462) | 1.8 | 0.3 | 8 | 23 | 68 | 1 | 25 | 75 |
| India | 211 | 16 | (194) | 1.4 | 0.2 | 1 | 1 | 97 | 16 | 74 | 10 |
| Indonesia | 208 | 1,189 | 981 | 1.7 | 6.4 | - | - | 100 | 3 | 97 | - |
| Iran | 263 | 0 | (263) | 2.3 | .0 | 8 | 37 | 55 | - | 100 | - |
| Iraq | 158 | - | (158) | 1.6 | 0.0 | 0 | 78 | 22 | - | - | - |
| Israel | 187 | 13 | (173) | 1.8 | 0.2 | 12 | 31 | 58 | - | 72 | 28 |
| Japan | 5,872 | 771 | (5,101) | 4.5 | 0.4 | 59 | 18 | 24 | 1 | 10 | 89 |
| Jordan | 51 | - | (51) | 1.9 | 0.0 | 16 | 48 | 36 | - | - | - |
| Kampuchea, Dem | 2 | 0 | (2) | NA | NA | - | - | 100 | 54 | 46 | - |
| Korea, Dem P R | 2 | - | (2) | NA | NA | 7 | 3 | 89 | - | - | - |
| Korea, Rep of | 855 | 132 | (723) | 2.7 | 0.4 | 57 | 3 | 39 | - | 47 | 53 |
| Kuwait | 114 | 23 | (91) | 1.7 | 0.2 | 1 | 68 | 31 | 3 | 67 | 30 |
| Lao PDR | - | 2 | 2 | 0.0 | 8.1 | - | - | - | 94 | 6 | - |
| Lebanon | 66 | 3 | (62) | 2.9 | 0.7 | 2 | 43 | 55 | - | 100 | - |
| Macao | 8 | 0 | (8) | NA | NA | 17 | 27 | 56 | - | - | 100 |
| Malaysia | 227 | 1,703 | 1,476 | 1.8 | 11.1 | - | 9 | 89 | 66 | 34 | 0 |
| Mongolia | 7 | 9 | 3 | NA | NA | - | 4 | 96 | 31 | 69 | - |
| Nepal | - | 12 | 12 | 0.0 | 7.5 | - | - | - | 100 | - | - |
| Oman | 56 | - | (56) | 1.8 | 0.0 | 4 | 85 | 11 | - | - | - |
| Pakistan | 104 | - | (104) | 1.8 | 0.0 | 2 | 20 | 79 | - | - | - |
| Philippines | 74 | 229 | 155 | 1.4 | 5.0 | - | 0 | 100 | 30 | 69 | 0 |
| Qatar | 18 | - | (18) | NA | NA | 9 | 37 | 54 | - | - | - |
| Saudi Arabia | 512 | - | (512) | 2.2 | 0.0 | 8 | 74 | 18 | - | - | - |
| Singapore | 363 | 419 | 55 | 1.4 | 1.8 | 7 | 41 | 51 | 3 | 76 | 21 |
| Sri Lanka | 25 | 1 | (24) | 1.4 | 0.1 | - | 17 | 83 | 100 | - | - |
| Syrian Arab Rep | 111 | 0 | (111) | 2.9 | .0 | 4 | 61 | 35 | 100 | - | - |
| Thailand | 253 | 39 | (214) | 2.7 | 0.6 | 17 | 16 | 68 | 8 | 67 | 25 |
| Turkey | 105 | 51 | (54) | 1.0 | 0.6 | 2 | 6 | 91 | 33 | 27 | 41 |
| Viet Nam | 9 | - | (9) | NA | NA | 39 | 31 | 30 | - | - | - |
| Yemen, PDR | 11 | 0 | (11) | 0.7 | .0 | 23 | 64 | 13 | - | 48 | - |
| Totals | 12,457 | 5,226 | (7,231) | 3.0 | 1.2 | 42 | 22 | 37 | 26 | 53 | 21 |

Table C.13 - The Structure of Trade in Forest Products, Europe, Oceania, USSR and World Totals

| | 1985 Trade in Forest Products | | | Forest Products in Total Trade | | Structure of 1985 Imports | | | Structure of 1985 Exports | | |
|---|---|---|---|---|---|---|---|---|---|---|---|
| | Imports | Exports | Net Exports | Imports | Exports | Logs | Mech Wood Products | Pulp & Paper | Logs | Mech Wood Products | Pulp & Paper |
| | .... (US$ million) ..... | | | .. (percent) .. | | ...... (percent) ....... | | | ...... (percent) ....... | | |
| **EUROPE** | | | | | | | | | | | |
| Albania | 0 | 1 | 0 | NA | NA | - | - | 100 | - | 82 | 18 |
| Austria | 620 | 1,311 | 691 | 3.0 | 7.7 | 24 | 19 | 57 | 4 | 39 | 57 |
| Belgium/Lux | 1,221 | 1,091 | (130) | 2.2 | 2.0 | 9 | 29 | 62 | 5 | 25 | 70 |
| Bulgaria | 157 | 22 | (135) | 1.1 | 0.2 | 10 | 15 | 75 | 3 | 45 | 52 |
| Czechoslovakia | 105 | 375 | 270 | 0.6 | 2.1 | 7 | 32 | 61 | 23 | 34 | 43 |
| Denmark | 885 | 186 | (699) | 4.9 | 1.1 | 2 | 45 | 53 | 13 | 34 | 53 |
| Finland | 298 | 4,604 | 4,306 | 2.3 | 33.8 | 58 | 7 | 35 | 2 | 20 | 78 |
| France | 3,014 | 1,578 | (1,436) | 2.8 | 1.6 | 6 | 18 | 75 | 13 | 20 | 67 |
| German Dem Rep | 461 | 113 | (347) | 1.8 | 0.4 | 6 | 53 | 41 | 27 | 3 | 71 |
| Germany, Fed Rep | 5,064 | 2,885 | (2,179) | 3.2 | 1.6 | 5 | 21 | 74 | 6 | 19 | 75 |
| Greece | 285 | 32 | (253) | 2.8 | 0.7 | 9 | 27 | 64 | 0 | 44 | 56 |
| Hungary | 255 | 85 | (170) | 3.1 | 1.0 | 15 | 31 | 54 | 49 | 22 | 29 |
| Iceland | 36 | - | (36) | NA | NA | 1 | 54 | 45 | - | - | - |
| Ireland | 270 | 33 | (236) | 2.7 | 0.3 | 1 | 33 | 66 | 15 | 57 | 28 |
| Italy | 2,653 | 720 | (1,933) | 2.9 | 0.9 | 14 | 32 | 54 | 0 | 21 | 79 |
| Malta | 17 | - | (17) | NA | NA | 1 | 41 | 58 | - | - | - |
| Netherlands | 1,872 | 863 | (1,010) | 2.9 | 1.3 | 2 | 37 | 61 | 2 | 12 | 86 |
| Norway | 432 | 808 | 376 | 2.8 | 4.1 | 15 | 40 | 45 | 2 | 6 | 91 |
| Poland | 175 | 154 | (21) | 1.6 | 1.3 | 3 | 36 | 61 | 34 | 42 | 24 |
| Portugal | 166 | 540 | 375 | 2.2 | 9.5 | 36 | 3 | 61 | 4 | 26 | 70 |
| Romania | 123 | 303 | 180 | 1.1 | 2.5 | 6 | 34 | 60 | 2 | 67 | 32 |
| Spain | 584 | 380 | (205) | 1.9 | 1.6 | 14 | 30 | 56 | 5 | 20 | 75 |
| Sweden | 557 | 4,930 | 4,373 | 2.0 | 16.2 | 33 | 24 | 44 | 1 | 24 | 75 |
| Switzerland | 744 | 379 | (365) | 2.4 | 1.4 | 5 | 29 | 66 | 9 | 19 | 72 |
| United Kingdom | 5,123 | 704 | (4,419) | 4.7 | 0.7 | 1 | 35 | 64 | 3 | 7 | 90 |
| Yugoslavia | 278 | 402 | 124 | 2.3 | 3.8 | 29 | 12 | 59 | 5 | 56 | 39 |
| Totals | 25,396 | 22,498 | (2,898) | 2.9 | 2.6 | 8 | 29 | 64 | 5 | 23 | 73 |
| **OCEANIA** | | | | | | | | | | | |
| Australia | 824 | 196 | (628) | 3.2 | 0.9 | 0 | 32 | 68 | 78 | 5 | 17 |
| Fiji | 8 | 6 | (2) | NA | NA | - | 7 | 93 | 4 | 96 | - |
| French Polynesia | 15 | - | (15) | NA | NA | 2 | 75 | 23 | - | - | - |
| New Caledonia | 9 | - | (9) | NA | NA | 8 | 78 | 14 | - | - | - |
| New Zealand | 114 | 305 | 191 | 1.9 | 5.3 | 1 | 15 | 84 | 12 | 25 | 64 |
| Papua New Guinea | 6 | 69 | 64 | 0.5 | 7.5 | - | - | 100 | 95 | 5 | - |
| Samoa (Western) | 5 | 1 | (3) | NA | NA | - | 42 | 58 | - | 100 | - |
| Solomon Islands | 2 | 19 | 17 | NA | NA | - | 100 | - | 95 | 5 | - |
| Tonga | 2 | - | (2) | NA | NA | - | 100 | - | - | - | - |
| Vanuatu | 0 | 1 | - | NA | NA | - | 100 | - | 81 | 19 | - |
| Totals | 984 | 597 | (386) | 3.0 | 2.0 | 0 | 31 | 69 | 46 | 16 | 38 |
| USSR | 889 | 2,726 | 1,837 | 1.1 | 3.1 | 4 | 13 | 83 | 27 | 42 | 31 |
| World Totals | 55,127 | 49,945 | 5,182 | 2.8 | 2.6 | 14 | 29 | 57 | 11 | 29 | 60 |

Table C.14 - Production Cost Comparison, Bleached Softwood Kraft Pulp

(in local currency and US$ per ton at 1985 prices)

| | US South | US Pacific Northwest | Canada BC Coast | Canada BC Interior | Eastern Canada | Sweden | Finland | New Zealand | Brazil | Chile |
|---|---|---|---|---|---|---|---|---|---|---|
| **Per-ton production costs in local currency:** | US$ | US$ | Cdn$ | Cdn$ | Cdn$ | Skr | Fmk | NZ$ | '000 Crz | Ch$ |
| Variable Costs | | | | | | | | | | |
| Wood | 137 | 133 | 179 | 154 | 195 | 1614 | 1397 | 175 | 530 | 11000 |
| Energy | 45 | 60 | 67 | 53 | 52 | 85 | 52 | 70 | 175 | 4600 |
| Chemicals | 41 | 37 | 53 | 52 | 66 | 242 | 219 | 77 | 350 | 9700 |
| Labor | 46 | 49 | 75 | 61 | 61 | 300 | 202 | 72 | 200 | 4100 |
| Other Materials | 44 | 42 | 59 | 51 | 46 | 225 | 172 | 71 | 240 | 4400 |
| - Sub-total | 313 | 321 | 433 | 371 | 420 | 2466 | 2042 | 465 | 1495 | 33800 |
| Fixed Costs | | | | | | | | | | |
| Overheads | 37 | 52 | 51 | 50 | 52 | 164 | 174 | 44 | 130 | 2800 |
| Depreciation | 28 | 29 | 33 | 37 | 33 | 155 | 139 | 38 | 490 | 11300 |
| Interest | 40 | 42 | 59 | 66 | 59 | 268 | 226 | 57 | 350 | 8300 |
| - Sub-total | 105 | 123 | 143 | 153 | 144 | 587 | 539 | 139 | 970 | 22400 |
| Total Cost at Mill Gate | 418 | 444 | 576 | 524 | 564 | 3053 | 2581 | 604 | 2465 | 56200 |
| **Per-ton production costs in US$ converted at:** | 1.00 | 1.00 | 1.37 | 1.37 | 1.37 | 8.54 | 6.02 | 1.74 | 7.60 | 180.00 |
| Variable Costs | | | | | | | | | | |
| Wood | 137 | 133 | 131 | 112 | 142 | 189 | 232 | 101 | 70 | 61 |
| Energy | 45 | 60 | 49 | 39 | 38 | 10 | 9 | 40 | 23 | 26 |
| Chemicals | 41 | 37 | 39 | 38 | 48 | 28 | 36 | 44 | 46 | 54 |
| Labor | 46 | 49 | 55 | 45 | 45 | 35 | 34 | 41 | 26 | 23 |
| Other Materials | 44 | 42 | 43 | 37 | 34 | 26 | 29 | 41 | 32 | 24 |
| - Sub-total | 313 | 321 | 316 | 271 | 307 | 289 | 339 | 267 | 197 | 188 |
| Fixed Costs | | | | | | | | | | |
| Overheads | 37 | 52 | 37 | 36 | 38 | 19 | 29 | 25 | 17 | 16 |
| Depreciation | 28 | 29 | 24 | 27 | 24 | 18 | 23 | 22 | 64 | 63 |
| Interest | 40 | 42 | 43 | 48 | 43 | 31 | 38 | 33 | 46 | 46 |
| - Sub-total | 105 | 123 | 104 | 112 | 105 | 69 | 90 | 80 | 128 | 124 |
| Total Cost at Mill Gate | 418 | 444 | 420 | 382 | 412 | 357 | 429 | 347 | 324 | 312 |
| **Freight in US$ per ton to:** | | | | | | | | | | |
| (a) West Europe | 60 | 50 | 50 | 65 | 45 | 25 | 30 | 65 | 45 | 55 |
| (b) Japan | 60 | 40 | 40 | 55 | 65 | 60 | 60 | 45 | 55 | 45 |
| **Delivered Cost to:** | | | | | | | | | | |
| (a) West Europe | 478 | 494 | 470 | 447 | 457 | 382 | 459 | 412 | 369 | 367 |
| (b) Japan | 478 | 484 | 460 | 437 | 477 | 417 | 489 | 392 | 379 | 357 |
| Approximate Shipments of Market Pulp in 1985 (thousand tons) | 1600 | 360 | 1700 | 1100 | 2500 | 2300 | 1100 | 90 | 100 | 270 |

Sources: Production cost figures for US, Canada and Scandanavia: Data Resources Inc. 'Pulp & Paper Review'. Lexington, Mass. August, 1985. Other data based on specific mill information and industry sources.

Table C.15 - Production Cost Comparisons, Newsprint

(in local currency and US$ per ton at 1985 prices)

| | US South | US Pacific Northwest | Canada BC Coast | Eastern Canada | Sweden | Finland | New Zealand | Chile |
|---|---|---|---|---|---|---|---|---|
| **Production costs in local currency:** | US$ | US$ | Cdn$ | Cdn$ | Skr | Fmk | NZ$ | Ch$ |
| Variable Costs | | | | | | | | |
| Wood | 77 | 64 | 99 | 146 | 1011 | 665 | 97 | 6000 |
| Energy | 111 | 121 | 101 | 78 | 534 | 449 | 120 | 12000 |
| Chemicals | 17 | 18 | 14 | 13 | 91 | 59 | 21 | 3000 |
| Labor | 74 | 91 | 112 | 111 | 369 | 316 | 90 | 10000 |
| Other Materials | 46 | 45 | 77 | 57 | 330 | 257 | 95 | 10000 |
| - Sub-total | 325 | 339 | 403 | 405 | 2335 | 1746 | 423 | 41000 |
| Fixed Costs | | | | | | | | |
| Overheads | 38 | 29 | 59 | 58 | 146 | 147 | 55 | 5400 |
| Capital Charges | 54 | 68 | 70 | 51 | 312 | 232 | 75 | 10800 |
| - Sub-total | 92 | 97 | 129 | 109 | 458 | 379 | 130 | 16200 |
| Total Cost at Mill Gate | 417 | 436 | 532 | 514 | 2793 | 2125 | 553 | 57200 |
| **Production costs in US$ converted at** | 1.00 | 1.00 | 1.35 | 1.35 | 8.54 | 6.02 | 1.74 | 180.00 |
| Variable Costs | | | | | | | | |
| Wood | 77 | 64 | 73 | 108 | 118 | 110 | 56 | 33 |
| Energy | 111 | 121 | 75 | 58 | 63 | 75 | 69 | 67 |
| Chemicals | 17 | 18 | 10 | 10 | 11 | 10 | 12 | 17 |
| Labor | 74 | 91 | 83 | 82 | 43 | 52 | 52 | 56 |
| Other Materials | 46 | 45 | 57 | 42 | 39 | 43 | 55 | 56 |
| - Sub-total | 325 | 339 | 299 | 300 | 273 | 290 | 243 | 228 |
| Fixed Costs | | | | | | | | |
| Overheads | 38 | 29 | 44 | 43 | 17 | 24 | 32 | 30 |
| Capital Charges | 54 | 68 | 52 | 38 | 37 | 39 | 43 | 60 |
| - Sub-total | 92 | 97 | 96 | 81 | 54 | 63 | 75 | 90 |
| Total Cost at Mill Gate | 417 | 436 | 394 | 381 | 327 | 353 | 318 | 318 |
| **Freight in US$ per ton to:** | | | | | | | | |
| (a) New York | 22 | 47 | 47 | 35 | NA | NA | NA | 55 |
| (b) Japan | 60 | 40 | 42 | NA | NA | NA | 45 | 55 |
| (c) West Europe | NA | NA | 50 | 45 | 25 | 30 | NA | 45 |
| Delivered Cost to: | | | | | | | | |
| (a) New York | 439 | 483 | 441 | 416 | NA | NA | NA | 373 |
| (b) Japan | 477 | 476 | 436 | NA | NA | NA | 363 | 373 |
| (c) West Europe | NA | NA | 444 | 426 | 352 | 383 | NA | 363 |
| Approximate Shipments of Newsprint in 1985 (thousand tons) | 2000 | 2000 | 4400 | 4400 | 1600 | 1800 | 300 | 200 |

Sources: Production cost figures for US, Canada and Scandanavia: Data Resources Inc. 'Pulp & Paper Review'. Lexington, Mass. August, 1985. Other data based on specific mill information and industry sources.

### Table C.16 - Average Import Prices of Forest Products to the United Kingdom

| | Coniferous Sawnwood | | | Plywood | | | Particleboard | | | Bleached Kraft Pulp | | | Newsprint | | |
|---|---|---|---|---|---|---|---|---|---|---|---|---|---|---|---|
| | current | constant 1986 L | index 1986=100 | current | constant 1986 L | index 1986=100 | current | constant 1986 L | index 1986=100 | current | constant 1986 L | index 1986=100 | current | constant 1986 L | index 1986=100 |
| | (UK L per m3) | | | (UK L per m3) | | | (UK L per m3) | | | (UK L per t) | | | (UK L per t) | | |
| 1965 | 19.0 | 74.0 | 71 | 53.0 | 206.4 | 97 | 22.6 | 88.0 | 73 | 54.0 | 210.3 | 70 | 52.0 | 202.5 | 59 |
| 1966 | 19.0 | 72.3 | 70 | 53.0 | 201.6 | 95 | 22.6 | 86.0 | 71 | 54.8 | 208.4 | 70 | 52.6 | 200.2 | 58 |
| 1967 | 18.3 | 68.0 | 66 | 51.7 | 192.2 | 90 | 22.4 | 83.3 | 69 | 56.1 | 208.6 | 70 | 54.3 | 201.7 | 59 |
| 1968 | 20.2 | 76.8 | 74 | 55.0 | 209.2 | 98 | 25.0 | 95.1 | 79 | 60.6 | 230.5 | 77 | 61.4 | 233.6 | 68 |
| 1969 | 22.1 | 80.3 | 78 | 60.8 | 221.1 | 104 | 27.5 | 100.0 | 83 | 64.0 | 232.6 | 78 | 62.3 | 226.6 | 66 |
| 1970 | 23.6 | 82.1 | 79 | 63.3 | 220.4 | 103 | 29.2 | 101.5 | 84 | 73.0 | 254.1 | 85 | 64.7 | 225.2 | 65 |
| 1971 | 24.7 | 82.5 | 80 | 64.5 | 215.3 | 101 | 30.8 | 102.9 | 85 | 76.1 | 254.1 | 85 | 75.3 | 251.3 | 73 |
| 1972 | 24.5 | 74.2 | 72 | 65.2 | 197.4 | 93 | 33.2 | 100.5 | 83 | 74.4 | 225.2 | 75 | 70.4 | 213.1 | 62 |
| 1973 | 36.3 | 90.0 | 87 | 89.7 | 222.4 | 104 | 41.2 | 102.1 | 85 | 88.1 | 218.3 | 73 | 82.0 | 203.2 | 59 |
| 1974 | 59.6 | 106.0 | 102 | 124.4 | 221.2 | 104 | 54.3 | 96.5 | 80 | 141.5 | 251.6 | 84 | 132.1 | 234.9 | 68 |
| 1975 | 55.3 | 90.5 | 87 | 113.9 | 186.3 | 87 | 62.6 | 102.3 | 85 | 186.8 | 305.6 | 102 | 161.1 | 263.5 | 77 |
| 1976 | 67.5 | 108.3 | 105 | 139.5 | 223.8 | 105 | 72.0 | 115.4 | 96 | 229.8 | 368.5 | 123 | 192.1 | 308.1 | 89 |
| 1977 | 81.8 | 120.5 | 116 | 170.2 | 250.8 | 118 | 82.5 | 121.6 | 101 | 224.6 | 331.0 | 111 | 235.0 | 346.3 | 101 |
| 1978 | 74.8 | 100.3 | 97 | 160.5 | 215.1 | 101 | 82.8 | 111.1 | 92 | 179.8 | 241.1 | 81 | 228.0 | 305.7 | 89 |
| 1979 | 82.1 | 92.0 | 89 | 178.6 | 200.2 | 94 | 85.8 | 96.1 | 80 | 207.0 | 231.9 | 78 | 226.0 | 253.2 | 74 |
| 1980 | 92.3 | 84.8 | 82 | 181.0 | 166.3 | 78 | 92.4 | 84.9 | 71 | 230.0 | 211.4 | 71 | 270.0 | 248.1 | 72 |
| 1981 | 90.3 | 83.8 | 81 | 190.3 | 176.7 | 83 | 85.8 | 79.6 | 66 | 269.0 | 249.7 | 84 | 283.0 | 262.7 | 76 |
| 1982 | 87.1 | 84.0 | 81 | 220.5 | 212.6 | 100 | 99.4 | 95.9 | 80 | 280.0 | 270.0 | 90 | 300.0 | 289.3 | 84 |
| 1983 | 100.9 | 103.0 | 99 | 212.9 | 217.4 | 102 | 101.5 | 103.7 | 86 | 275.0 | 280.8 | 94 | 300.0 | 306.3 | 89 |
| 1984 | 116.0 | 120.9 | 117 | 232.7 | 242.5 | 114 | 104.0 | 108.4 | 90 | 378.0 | 393.9 | 132 | 310.0 | 323.0 | 94 |
| 1985 | 107.7 | 114.3 | 110 | 229.1 | 243.2 | 114 | 102.8 | 109.1 | 91 | 301.2 | 319.6 | 107 | 333.1 | 353.5 | 103 |
| 1986 | 103.5 | 103.5 | 100 | 213.2 | 213.2 | 100 | 120.4 | 120.4 | 100 | 298.6 | 298.6 | 100 | 344.3 | 344.3 | 100 |
| Trend Line Annual Average Increase | | | 1.8% | | | none | | | none | | | 1.8% | | | 2.4% |

Notes: (1) Current prices are average unit prices of imports from all sources in UK pounds per unit.
(2) Constant prices have been obtained by deflating current prices with IMF's Index of Export Unit Values.

Table C.17 - Regional Breakdown of Median Projection of Demand for Forest Products (2000)

| | Sawnwood | | | | Wood-based Panels | | | | | | Paper/Paperboard | | | | | |
|---|---|---|---|---|---|---|---|---|---|---|---|---|---|---|---|---|
| | Hardwood | | Softwood | | Plywood & Veneer | | Particle-board | | Fiberboard | | Newsprint | | Printing & Writing | | Other Grades | |
| | 1985 | 2000 | 1985 | 2000 | 1985 | 2000 | 1985 | 2000 | 1985 | 2000 | 1985 | 2000 | 1985 | 2000 | 1985 | 2000 |
| | ................(millions of cubic meter ............... | | | | | | | | | | ......... (million tons) ......... | | | | | |
| AFRICA | | | | | | | | | | | | | | | | |
| Northern Africa | 0.4 | 0.7 | 3.7 | 7.7 | 0.5 | 1.2 | 0.2 | 0.4 | .0 | .0 | 0.1 | 0.1 | 0.3 | 0.5 | 0.7 | 1.3 |
| South Africa | 0.5 | 0.7 | 1.3 | 1.5 | 0.1 | 0.2 | 0.2 | 0.3 | 0.1 | 0.1 | 0.2 | 0.3 | 0.3 | 0.5 | 0.9 | 1.1 |
| Other Africa | 4.9 | 6.1 | 0.4 | 0.5 | 0.7 | 0.9 | 0.1 | 0.1 | 0.1 | 0.1 | 0.1 | 0.1 | 0.1 | 0.2 | 0.3 | 0.3 |
| Regional Total | 5.7 | 7.5 | 5.4 | 9.8 | 1.3 | 2.2 | 0.5 | 0.8 | 0.2 | 0.2 | 0.4 | 0.5 | 0.7 | 1.1 | 1.9 | 2.7 |
| ASIA | | | | | | | | | | | | | | | | |
| Japan | 5.6 | 5.4 | 28.0 | 28.0 | 5.9 | 6.2 | 1.3 | 2.1 | 0.6 | 0.5 | 2.8 | 3.9 | 4.6 | 8.5 | 13.1 | 16.6 |
| China | 10.6 | 19.5 | 17.2 | 31.6 | 1.7 | 4.6 | 0.3 | 0.8 | 0.8 | 1.7 | 0.7 | 1.2 | 2.8 | 6.4 | 7.8 | 17.6 |
| Southeast Asia | 11.3 | 19.7 | .0 | 0.1 | 2.3 | 6.3 | 0.1 | 0.4 | 0.1 | 0.3 | 0.5 | 0.9 | 0.6 | 1.3 | 1.4 | 3.1 |
| Middle East | 1.5 | 1.9 | 6.2 | 8.0 | 1.1 | 1.7 | 0.8 | 1.2 | 0.2 | 0.2 | 0.3 | 0.4 | 0.4 | 0.5 | 0.7 | 1.0 |
| Other Asia | 16.8 | 29.3 | 5.0 | 9.2 | 1.8 | 4.8 | 0.3 | 0.9 | 0.1 | 0.2 | 0.9 | 1.8 | 1.8 | 4.0 | 3.0 | 6.8 |
| Regional Total | 45.8 | 75.8 | 56.5 | 77.1 | 12.9 | 23.6 | 2.7 | 5.4 | 1.8 | 3.0 | 5.2 | 8.1 | 10.1 | 20.8 | 26.0 | 45.1 |
| NORTH AMERICA | | | | | | | | | | | | | | | | |
| Canada | 1.7 | 1.7 | 13.8 | 15.1 | 2.0 | 2.2 | 1.5 | 3.0 | 0.7 | 0.6 | 1.1 | 1.5 | 1.7 | 3.2 | 2.5 | 3.2 |
| USA | 13.1 | 12.0 | 103.6 | 111.1 | 20.9 | 23.1 | 8.4 | 16.4 | 5.3 | 4.5 | 12.2 | 17.3 | 18.5 | 35.1 | 38.4 | 48.9 |
| Regional Total | 14.8 | 13.7 | 117.5 | 126.2 | 22.9 | 25.3 | 9.9 | 19.4 | 6.0 | 5.1 | 13.3 | 18.8 | 20.2 | 38.3 | 41.0 | 52.1 |
| LATIN AMERICA | 13.9 | 21.6 | 12.6 | 18.0 | 1.9 | 2.9 | 1.8 | 4.3 | 0.8 | 1.6 | 1.3 | 2.0 | 2.4 | 3.8 | 6.7 | 12.7 |
| WESTERN EUROPE | | | | | | | | | | | | | | | | |
| Scandanavia | 0.9 | 0.8 | 10.7 | 10.7 | 0.6 | 0.6 | 2.0 | 2.6 | 0.6 | 0.5 | 0.8 | 1.0 | 1.5 | 2.8 | 2.4 | 3.1 |
| Other | 14.6 | 14.0 | 44.7 | 44.7 | 5.3 | 5.9 | 16.6 | 21.5 | 2.1 | 1.8 | 5.4 | 7.5 | 13.7 | 25.4 | 23.2 | 29.5 |
| Regional Total | 15.5 | 14.8 | 55.4 | 55.4 | 5.9 | 6.5 | 18.6 | 24.1 | 2.7 | 2.3 | 6.2 | 8.6 | 15.2 | 28.2 | 25.6 | 32.6 |
| EASTERN EUROPE & USSR | | | | | | | | | | | | | | | | |
| Eastern Europe | 6.6 | 6.2 | 16.2 | 16.2 | 1.2 | 1.2 | 5.1 | 8.6 | 1.5 | 1.3 | 0.5 | 0.8 | 1.0 | 1.6 | 4.4 | 5.4 |
| USSR | 11.4 | 10.6 | 79.3 | 79.3 | 2.7 | 2.7 | 6.6 | 11.3 | 3.1 | 2.7 | 1.2 | 1.7 | 1.6 | 2.6 | 7.0 | 8.6 |
| Regional Total | 18.0 | 16.8 | 95.5 | 95.5 | 3.9 | 3.9 | 11.7 | 19.9 | 4.6 | 4.0 | 1.7 | 2.5 | 2.5 | 4.2 | 11.5 | 14.0 |
| OCEANIA | | | | | | | | | | | | | | | | |
| Australia/NZ | 2.3 | 2.2 | 3.8 | 3.8 | 0.3 | 0.3 | 0.8 | 1.3 | 0.2 | 0.2 | 0.8 | 1.0 | 0.5 | 0.9 | 1.5 | 1.8 |
| Other Oceania | 0.2 | 0.2 | 0.1 | 0.2 | .0 | .0 | .0 | .0 | .0 | .0 | .0 | .0 | .0 | .0 | .0 | .0 |
| Regional Total | 2.5 | 2.4 | 4.0 | 4.1 | 0.3 | 0.3 | 0.8 | 1.3 | 0.2 | 0.2 | 0.8 | 1.1 | 0.5 | 0.9 | 1.5 | 1.8 |
| WORLD | | | | | | | | | | | | | | | | |
| Developing | 59.4 | 99.0 | 45.3 | 75.5 | 10.0 | 22.4 | 3.5 | 8.2 | 2.1 | 4.2 | 3.8 | 6.4 | 8.3 | 16.7 | 20.5 | 42.8 |
| Industrial | 56.7 | 53.5 | 301.5 | 310.5 | 39.0 | 42.2 | 42.5 | 67.1 | 14.2 | 12.3 | 25.1 | 35.1 | 43.3 | 80.7 | 93.5 | 118.2 |
| World Total | 116.1 | 152.6 | 346.9 | 386.0 | 49.0 | 64.7 | 45.9 | 75.3 | 16.3 | 16.5 | 28.9 | 41.5 | 51.7 | 97.4 | 114.0 | 161.1 |

Note: 1985 figures are averages, 1984-86

Table C.18 - World Balance for Tropical Hardwoods and Tropical Hardwood Products (1985 and 2000)

| | Total Log Cut | | Domestic Production | | | | Domestic Consumption | | | | Exports | | | | | |
|---|---|---|---|---|---|---|---|---|---|---|---|---|---|---|---|---|
| | | | Sawnwood | | Plywood | | Sawnwood | | Plywood | | Logs | | Sawnwood | | Plywood | |
| | 1985 | 2000 | 1985 | 2000 | 1985 | 2000 | 1985 | 2000 | 1985 | 2000 | 1985 | 2000 | 1985 | 2000 | 1985 | 2000 |
| | (millions of cubic meters) | | | | | | | | | | | | | | | |
| ASIA | | | | | | | | | | | | | | | | |
| Indonesia | 23500 | 37400 | 7065 | 11000 | 4615 | 7000 | 4899 | 6600 | 932 | 1500 | 300 | - | 2166 | 4400 | 3683 | 5500 |
| Malaysia | 30880 | 24000 | 7138 | 8000 | 1429 | 1500 | 4398 | 5900 | 651 | 1000 | 19771 | 4700 | 2740 | 2100 | 778 | 500 |
| Philippines | 4206 | 3880 | 1200 | 1400 | 586 | 400 | 693 | 900 | 248 | 400 | 679 | 200 | 507 | 500 | 338 | - |
| Thailand | 1817 | 1000 | 909 | 1000 | 223 | 200 | 909 | 1200 | 223 | 300 | - | - | - | - | - | - |
| Other | 2100 | 1100 | 500 | 500 | - | - | 400 | 500 | - | - | 100 | 100 | 100 | - | 80 | - |
| Sub-total | 62503 | 67380 | 16812 | 21900 | 6853 | 9100 | 11299 | 15100 | 2054 | 3200 | 20850 | 5000 | 5513 | 7000 | 4879 | 6000 |
| AFRICA | | | | | | | | | | | | | | | | |
| Cameroon | 2093 | 1930 | 565 | 500 | 73 | 150 | 479 | 560 | 36 | 50 | 746 | 600 | 86 | - | 37 | 100 |
| Cent Afr Rep | 269 | 344 | 56 | 100 | 7 | 20 | 24 | 30 | 3 | 4 | 65 | 100 | 32 | 70 | 4 | 16 |
| Congo | 572 | 720 | 50 | 100 | 74 | 100 | 29 | 30 | 24 | 30 | 275 | 300 | 21 | 70 | 50 | 70 |
| Gabon | 1484 | 2960 | 106 | 150 | 228 | 300 | 105 | 120 | 174 | 230 | 1089 | 2000 | 1 | 30 | 54 | 70 |
| Ivory Coast | 3895 | 3430 | 646 | 1200 | 148 | 150 | 232 | 270 | 63 | 80 | 1345 | 700 | 414 | 930 | 85 | 70 |
| Nigeria | 5700 | 6620 | 2700 | 3200 | 154 | 100 | 2698 | 3130 | 154 | 210 | - | - | 2 | 70 | - | - |
| Other | 1300 | 3320 | 350 | 1000 | 100 | 100 | 250 | 290 | 60 | 80 | 400 | 1100 | 100 | 710 | 40 | 20 |
| Sub-total | 15313 | 19324 | 4473 | 6250 | 784 | 920 | 3817 | 4430 | 514 | 684 | 3920 | 4800 | 656 | 1880 | 270 | 346 |
| LATIN AMERICA | | | | | | | | | | | | | | | | |
| Brazil | 14285 | 25520 | 8377 | 11000 | 1136 | 1600 | 7956 | 9200 | 855 | 1200 | 26 | - | 421 | 1800 | 281 | 400 |
| Colombia | 1900 | 1620 | 680 | 700 | 55 | 100 | 678 | 790 | 55 | 70 | - | - | 2 | - | - | 30 |
| Ecuador | 2350 | 3820 | 1213 | 1800 | 68 | 100 | 1181 | 1370 | 57 | 80 | - | - | 32 | 430 | 11 | 20 |
| Other | 1200 | 820 | 120 | 300 | 60 | 100 | 20 | 20 | 0 | 0 | 20 | - | 100 | 280 | 60 | 100 |
| Sub-total | 19735 | 31780 | 10390 | 13800 | 1319 | 1900 | 9835 | 11380 | 967 | 1350 | 46 | - | 555 | 2510 | 352 | 550 |
| OCEANIA | 2200 | 2110 | 120 | 250 | 21 | 50 | 95 | 110 | 10 | 13 | 1476 | 1500 | 25 | 140 | 11 | 37 |
| WORLD TOTALS | 99751 | 120594 | 31795 | 42200 | 8977 | 11970 | 25046 | 31020 | 3545 | 5247 | 26292 | 11300 | 6749 | 11530 | 5512 | 6933 |

Notes (1) Balances based on following conversion factors of logs to product:
Sawnwood 2.0 cubic meters
Plywood 2.2 cubic meters
(2) Only those countries whose principal non-coniferous log output is tropical hardwoods have been included

## Table C.19 - Self Sufficiency Trends in the Supply of Forest Products

| | Production | | | Consumption | | | Regional Self-Sufficiency (%) | | |
|---|---|---|---|---|---|---|---|---|---|
| | 1975 | 1985 | 2000 | 1975 | 1985 | 2000 | 1975 | 1985 | 2000 |
| **SAWNWOOD (million m3)** | | | | | | | | | |
| Industrial Countries | | | | | | | | | |
| - North America | 113.0 | 139.3 | 156 | 107.6 | 132.3 | 140 | 105% | 105% | 112% |
| - Western Europe | 58.7 | 65.1 | 67 | 66.7 | 70.9 | 70 | 88% | 92% | 95% |
| - E. Europe/USSR | 136.5 | 119.1 | 118 | 129.0 | 113.4 | 112 | 106% | 105% | 105% |
| - Japan | 38.9 | 28.4 | 28 | 41.9 | 33.7 | 33 | 93% | 84% | 85% |
| - Other | 7.2 | 6.8 | 7 | 8.4 | 7.9 | 8 | 85% | 86% | 90% |
| - sub-total | 354.2 | 358.7 | 377 | 353.6 | 358.2 | 364 | 100% | 100% | 103% |
| Developing Countries | | | | | | | | | |
| - Africa | 3.7 | 6.2 | 11 | 4.6 | 9.3 | 15 | 80% | 66% | 70% |
| - Asia & Oceania | 42.0 | 71.3 | 108 | 40.6 | 69.0 | 120 | 103% | 103% | 90% |
| - Latin America | 18.5 | 26.4 | 44 | 18.6 | 26.5 | 40 | 99% | 99% | 110% |
| - sub-total | 64.2 | 103.8 | 162 | 63.9 | 104.8 | 175 | 101% | 99% | 93% |
| World Total | 418.4 | 462.5 | 539 | 417.4 | 463.0 | 539 | 100% | 100% | 100% |
| **PANELS (million m3)** | | | | | | | | | |
| Industrial Countries | | | | | | | | | |
| - North America | 31.2 | 36.9 | 50 | 32.6 | 38.8 | 50 | 96% | 95% | 100% |
| - Western Europe | 24.1 | 24.1 | 30 | 25.2 | 27.2 | 33 | 95% | 89% | 91% |
| - E. Europe/USSR | 14.7 | 20.6 | 28 | 14.4 | 20.2 | 28 | 102% | 102% | 100% |
| - Japan | 8.5 | 7.2 | 8 | 8.7 | 7.8 | 9 | 97% | 92% | 95% |
| - Other | 1.4 | 1.6 | 2 | 1.4 | 1.7 | 2 | 96% | 97% | 100% |
| - sub-total | 79.8 | 90.4 | 118 | 82.3 | 95.6 | 122 | 97% | 95% | 97% |
| Developing Countries | | | | | | | | | |
| - Africa | 0.7 | 1.4 | 3 | 0.8 | 1.5 | 3 | 94% | 91% | 105% |
| - Asia | 5.8 | 13.5 | 27 | 3.2 | 9.6 | 23 | 180% | 140% | 114% |
| - Latin America | 2.8 | 4.8 | 9 | 3.2 | 4.5 | 9 | 90% | 108% | 100% |
| - sub-total | 9.4 | 19.7 | 38 | 7.2 | 15.6 | 35 | 131% | 126% | 110% |
| World Total | 89.2 | 110.1 | 156 | 89.5 | 111.2 | 156 | 100% | 99% | 100% |
| **PAPER AND PAPERBOARD (million tons)** | | | | | | | | | |
| Industrial Countries | | | | | | | | | |
| - North America | 61.5 | 76.6 | 117 | 57.5 | 74.4 | 109 | 107% | 103% | 107% |
| - Western Europe | 37.8 | 50.2 | 73 | 36.7 | 47.0 | 69 | 103% | 107% | 105% |
| - E. Europe/USSR | 13.5 | 15.7 | 21 | 13.8 | 15.7 | 21 | 97% | 100% | 100% |
| - Japan | 14.9 | 20.5 | 29 | 14.5 | 20.5 | 29 | 103% | 100% | 100% |
| - Other | 2.6 | 3.8 | 6 | 3.1 | 4.3 | 6 | 84% | 90% | 100% |
| - sub-total | 130.3 | 166.8 | 245 | 125.6 | 161.9 | 234 | 104% | 103% | 105% |
| Developing Countries | | | | | | | | | |
| - Africa | 0.3 | 0.6 | 1 | 0.9 | 1.5 | 2 | 35% | 38% | 40% |
| - Asia | 7.3 | 17.3 | 36 | 9.2 | 20.8 | 45 | 80% | 83% | 80% |
| - Latin America | 5.1 | 9.3 | 18 | 6.8 | 10.3 | 19 | 75% | 91% | 95% |
| - sub-total | 12.7 | 27.2 | 55 | 16.9 | 32.6 | 66 | 75% | 83% | 83% |
| World Total | 143.0 | 194.0 | 300 | 142.4 | 194.5 | 300 | 100% | 100% | 100% |

Table C.20 - Preliminary Estimate of Investment Requirements for Forest Industries

| | Total New Capacity Required (1985-2000 - millions/yr) | | | | Average Annual Investment Required | | | | | Total Gross Domestic Investment in 1985 | Annual Forest Investment as percent of Total Investment |
|---|---|---|---|---|---|---|---|---|---|---|---|
| | Sawnwood | Panels | Pulp | Paper | Sawnwood | Panels | Pulp | Paper | Total | | |
| | (cubic meters) | | .. (tons) .. | | .......... ($ million/year) .......... | | | | | (US$ billion) | |
| Industrial Countries | | | | | | | | | | | |
| - North America | 8.0 | 12.9 | 31.5 | 40.7 | 160 | 170 | 2,730 | 1,950 | 5,010 | 759 | 0.7% |
| - Western Europe | - | 5.9 | 11.4 | 22.5 | - | 80 | 990 | 1,080 | 2,150 | 622 | 0.3% |
| - E. Europe/USSR | - | 7.2 | 3.5 | 5.0 | - | 100 | 310 | 240 | 650 | 192 e | 0.3% |
| - Japan | - | 1.2 | 6.1 | 8.6 | - | 20 | 520 | 410 | 950 | 357 | 0.3% |
| - Other | - | 0.7 | 1.3 | 1.8 | - | 10 | 110 | 90 | 210 | 66 | 0.3% |
| - sub-total | 8.0 | 27.9 | 50.1 | 78.7 | 160 | 380 | 4,660 | 3,770 | 8,970 | 1,995 | 0.4% |
| Developing Countries | | | | | | | | | | | |
| - Africa | 5.5 | 1.5 | 0.4 | 0.5 | 110 | 20 | 30 | 20 | 180 | 65 | 0.3% |
| - Asia & Oceania | 23.4 | 11.5 | 16.4 | 23.4 | 470 | 150 | 1,420 | 1,120 | 3,160 | 330 | 1.0% |
| - Latin America | 12.9 | 4.0 | 12.6 | 10.3 | 260 | 50 | 1,090 | 490 | 1,890 | 116 | 1.6% |
| - sub-total | 41.8 | 16.9 | 28.9 | 34.2 | 840 | 220 | 2,540 | 1,630 | 5,230 | 511 | 1.0% |
| World Total | 49.8 | 44.8 | 79.0 | 112.9 | 1,000 | 600 | 7,200 | 5,400 | 14,200 | 2,506 | 0.6% |

Notes: (1) New capacity requirements are based on production projections in Table C.19, with some discounting to take account of present underutilized capacity for sawnwood.
(2) Unit costs of new capacity take into account regional differences in likely mill types and sizes.
(3) Average annual investment assumes that investments are spread evenly over the fifteen-year period to 2000.
(4) Average unit costs of new capacity:

| | |
|---|---|
| Sawnwood | $300 per annual cubic meter |
| Panels | $200 per annual cubic meter |
| Pulp | $1,300 per annual ton |
| Paper | $720 per annual ton |

# DISTRIBUTORS OF WORLD BANK PUBLICATIONS

**ARGENTINA**
Carlos Hirsch, SRL
Galeria Guemes
Florida 165, 4th Floor-Ofc. 453/465
1333 Buenos Aires

**AUSTRALIA, PAPUA NEW GUINEA, FIJI, SOLOMON ISLANDS, VANUATU, AND WESTERN SAMOA**
Info-Line
Overseas Document Delivery
Box 506, GPO
Sydney, NSW 2001

**AUSTRIA**
Gerold and Co.
A-1011 Wien
Graben 31

**BAHRAIN**
MEMBR Information Services
P.O. Box 2750
Manama Town 317

**BANGLADESH**
Micro Industries Development Assistance Society (MIDAS)
G.P.O. Box 800
Dhaka

**BELGIUM**
Publications des Nations Unies
Av. du Roi 202
1060 Brussels

**BRAZIL**
Publicacoes Tecnicas Internacionais Ltda.
Rua Peixoto Gomide, 209
01409 Sao Paulo, SP

**CANADA**
Le Diffuseur
C.P. 85, 1501 Ampere Street
Boucherville, Quebec
J4B 5E6

**COLOMBIA**
Enlace Ltda.
Carrera 6 No. 51-21
Bogota D.E.

Apartado Aereo 4430
Cali, Valle

**COSTA RICA**
Libreria Trejos
Calle 11-13
Av. Fernandez Guell
San Jose

**COTE D'IVOIRE**
Entre d'Edition et de Diffusion Africaines (CEDA)
04 B.P. 541
Abidjan 04 Plateau

**CYPRUS**
MEMRB Information Services
P.O. Box 2098
Nicosia

**DENMARK**
SamfundsLitteratur
Rosenoerns Alle 11
DK-1970 Frederiksberg C.

**DOMINICAN REPUBLIC**
Editora Taller, C. por A.
Restauracion
Apdo. postal 2190
Santo Domingo

**EGYPT, ARAB REPUBLIC OF**
Al Ahram
Al Galaa Street
Cairo

**FINLAND**
Akateeminen Kirjakauppa
P.O. Box 128
SF-00101
Helsinki 10

**FRANCE**
World Bank Publications
66, avenue d'Iéna
75116 Paris

**GERMANY, FEDERAL REPUBLIC OF**
UNO-Verlag
Poppelsdorfer Alle 55
D-5300 Bonn 1

**GREECE**
KEME
24, Ippodamou Street
Athens-11635

**GUATEMALA**
Librerias Piedra Santa
Centro Cultural Piedra Santa
11 calle 6-50 zona 1
Guatemala City

**HONG KONG, MACAU**
Asia 2000 Ltd.
6 Fl., 146 Prince Edward Road, W,
Kowloon
Hong Kong

**HUNGARY**
Kultura
P.O. Box 139
1389 Budapest 62

**INDIA**
Allied Publishers Private Ltd.
751 Mount Road
Madras—600 002

15 J.N. Heredia Marg
Ballard Estate
Bombay—400 038

13/14 Asaf Ali Road
New Delhi—110 002

17 Chittaranjan Avenue
Calcutta—700 072

Jayadeva Hostel Building
5th Main Road Gandhinagar
Bangalore—560 009

3-5-1129 Kachiguda Cross Road
Hyderabad—500 027

Prarthana Flats, 2nd Floor
Near Thakore Baug, Navrangpura
Ahmedabad—380 009

Patiala House
16-A Ashok Marg
Lucknow—226 001

**INDONESIA**
Pt. Indira Limited
Jl. Sam Ratulangi 37
Jakarta Pusat
P.O. Box 181

**IRELAND**
TDC Publishers
12 North Frederick Street
Dublin 1

**ISRAEL**
The Jerusalem Post
The Jerusalem Post Building
P.O. Box 81
Romema Jerusalem 91000

**ITALY**
Licosa Commissionaria Sansoni SPA
Via Lamarmora 45
Casella Postale 552
50121 Florence

**JAPAN**
Eastern Book Service
37-3, Hongo 3-Chome, Bunkyo-ku 113
Tokyo

**JORDAN**
Jordan Center for Marketing Research
P.O. Box 3143
Jabal
Amman

**KENYA**
Africa Book Service (E.A.) Ltd.
P.O. Box 45245
Nairobi

**KOREA, REPUBLIC OF**
Pan Korea Book Corporation
P.O. Box 101, Kwangwhamun
Seoul

**KUWAIT**
MEMRB
P.O. Box 5465

**MALAYSIA**
University of Malaya Cooperative Bookshop Limited
P.O. Box 1127, Jalan Pantai Baru
Kuala Lumpur

**MEXICO**
INFOTEC
Apartado Postal 22-860
Col. PE/A Pobre
14060 Tlalpan, Mexico D.F.

**MOROCCO**
Societe d'Etudes Marketing Marocaine
2 Rue Moliere, Bd. d'Anfa
Casablanca

**THE NETHERLANDS**
InOr Publikaties
Noorderwal 38
7241 BL Lochem

**NEW ZEALAND**
Hills Library and Information Service
Private Bag
New Market
Auckland

**NIGERIA**
University Press Limited
Three Crowns Building Jericho
Private Mail Bag 5095
Ibadan

**NORWAY**
Tanum-Karl Johan, A.S.
P.O. Box 1177 Sentrum
Oslo 1

**OMAN**
MEMRB Information Services
P.O. Box 1613, Seeb Airport
Muscat

**PAKISTAN**
Mirza Book Agency
65, Shahrah-e-Quaid-e-Azam
P.O. Box No. 729
Lahore 3

**PERU**
Editorial Desarrollo SA
Apartado 3824
Lima

**THE PHILIPPINES**
National Book Store
701 Royal Avenue
Metro Manila

**POLAND**
ORPAN
Palac Kultury i Nauki
00-901 WARSZAWA

**PORTUGAL**
Liveria Portugal
Rua Do Carmo 70-74
1200 Lisbon

**SAUDI ARABIA, QATAR**
Jarir Book Store
P.O. Box 3196
Riyadh 11471

**SINGAPORE, TAIWAN, BURMA, BRUNEI**
Information Publications
Private, Ltd.
02-06 1st Fl., Pei-Fu Industrial
Bldg., 24 New Industrial Road
Singapore

**SOUTH AFRICA**
*For single titles:*
Oxford University Press Southern Africa
P.O. Box 1141
Cape Town 8000

*For subscription orders:*
International Subscription Service
P.O. Box 41095
Craighall
Johannesburg 2024

**SPAIN**
Mundi-Prensa Libros, S.A.
Castello 37
28001 Madrid

**SRI LANKA AND THE MALDIVES**
Lake House Bookshop
P.O. Box 244
100, Sir Chittampalam A. Gardiner Mawatha
Colombo 2

**SWEDEN**
*For single titles:*
ABCE Fritzes Kungl. Hovbokhandel
Regeringsgatan 12, Box 16356
S-103 27 Stockholm

*For subscription orders:*
Wennergren-Williams AB
Box 30004
S-104 25 Stockholm

**SWITZERLAND**
Librairie Payot
6 Rue Grenus
Case postal 381
CH 1211 Geneva 11

**TANZANIA**
Oxford University Press
P.O. Box 5299
Dar es Salaam

**THAILAND**
Central Department Store
306 Silom Road
Bangkok

**TRINIDAD & TOBAGO, ANTIGUA, BARBUDA, BARBADOS, DOMINICA, GRENADA, GUYANA, JAMAICA, MONTSERRAT, ST. KITTS AND NEVIS, ST. LUCIA, ST. VINCENT & GRENADINES**
Systematics Studies Unit
55 Eastern Main Road
Curepe
Trinidad, West Indies

**TURKEY**
Haset Kitapevi A.S.
469, Istiklal Caddesi
Beyoglu-Istanbul

**UGANDA**
Uganda Bookshop
P.O. Box 7145
Kampala

**UNITED ARAB EMIRATES**
MEMBR Gulf Co.
P.O. Box 6097
Sharjah

**UNITED KINGDOM**
Microinfo Ltd.
P.O. Box 3
Alton, Hampshire GU 34 2PG
England

**VENEZUELA**
Libreria del Este
Aptdo. 60.337
Caracas 1060-A

**YUGOSLAVIA**
Jugoslovenska Knjiga
YU-11000 Belgrade Trg Republike

**ZIMBABWE**
Textbook Sales Pvt. Ltd.
Box 3799
Harare